THE NEW
RULES
of the
ROOST

THE NEW
RULES
of the
ROOST

. . . .

Organic Care & Feeding for the
FAMILY FLOCK

ROBERT AND HANNAH LITT

TIMBER PRESS
PORTLAND, OREGON

Published in 2018 by Timber Press, Inc.

The Haseltine Building
133 S.W. Second Avenue, Suite 450
Portland, Oregon 97204-3527
timberpress.com

Printed in China

Text design by Meredith Nelson
Cover design by Jarrod Taylor and Sarah Crumb

Library of Congress Cataloging-in-Publication Data

Names: Litt, Robert, author. | Litt, Hannah, author.
Title: The new rules of the roost: organic care & feeding for
the family flock / Robert Litt and Hannah Litt.
Description: Portland, Oregon: Timber Press, 2018. | Includes
bibliographical references and index. |
Identifiers: LCCN 2018001915 (print) | LCCN 2018004016 (ebook) |
ISBN 9781604698664 | ISBN 9781604698459 (hardcover) | ISBN 9781604698183 (pbk.)
Subjects: LCSH: Chickens. | Poultry farms. | Organic farming.
Classification: LCC SF487 (ebook) | LCC SF487 .L7753 2018 (print) | DDC 636.5—dc23
LC record available at https://lccn.loc.gov/2018001915

A catalog record for this book is also available from the British Library.

For our daughters, Abby and Mira—

*feisty, brilliant, beautiful—we love each of your
many facets. Thank you for sharing us with this book.
You may have your parents back now.*

CONTENTS

INTRODUCTION

Kauai, the oldest and northernmost of the Hawaiian Islands, is renowned for its rugged canyons, tropical beaches, and laid-back lifestyle. It also provided the lush jungle setting for the dinosaur epic *Jurassic Park*, in which the ancient beasts are brought back into the modern world through genetic science—with predictably chaotic results.

In reality, Kauai is teeming with living dinosaur relatives of *Tyrannosaurus rex*, and there is indeed a very real, and uncontrolled, genetic experiment underway. The tiny raptors are *Gallus gallus domesticus*, better known as chickens, and the island is overrun with them. Visitors are charmed by the antics of these familiar birds in this unexpected milieu, eagerly spotting them juking through the jungle greenery or cautiously collecting cabana crumbs. Locals regard them with disdain. They patiently explain that these are nothing more than fowl gone feral, an introduced irritation, escapees from local farms run amok in a relatively predator-free environment.

Local lore has it that island chickens escaped from their coops and cages in the chaos of Hurricane Iwa in 1982 and again with Hurricane Iniki in 1992. These free-ranging chickens began to breed with the local Polynesian birds of Indian origin. Subject to the forces of natural selection, hundreds of years of purposeful breeding were undone in a few generations. Now absent from the island are the fluffy Orpington, the noble Rhode Island Red, and the green-egg-laying Ameraucana. In their place thrives a diverse population of chickens that share both domestic and wild DNA, a comical rabble of recognizable farm animals and exotic avian majesty.

As so-called chicken-keeping experts, we were stunned and humbled by the spectacle of chickens by the thousands surviving, and indeed thriving, without the least bit of human care. Though we knew to expect the island birds, our first encounter with them was nothing less than a revelation, and seeing our first truly free-range, wild-breeding chicken on the lawn at the airport was a thrill (we're easily thrilled). She was small, perhaps 3 pounds, sporting brindle plumage that flashed iridescent green in the low morning sun. We set down our bags and approached for a gleeful gawk. She stopped feeding on some invisible tidbit in the grass and turned her head sideways to take a skeptical glance at us. We silently admired her for a moment, noting how she resembled her wild ancestor—the jungle fowl of Southeast Asia. These forest-dwelling, low-flying birds were the genetic source material for today's domestic chicken. This particular fowl was not in the least interested in our genetics, however, beyond identifying that we were some sort of large, chicken-devouring creatures with ugly feathers. We loudly fumbled for a camera, confirming for the small hen that we had some sort of malicious intentions, and she was gone in a tiny flash of green. We decided to name her Cluck.

At that very moment, back home on the mainland, our motley crew of plump, waddling hens would have been receiving their first visit from our well-paid chicken sitter. Our soft pets were no doubt abusing him with squawks about the perceived delay in food service or the lack of expected treats. "And another thing, whoever you are! The water is too cold! Please warm it before you go!" we imagined that Checkers, our constantly cranky boss hen, would have been demanding.

Cluck needs no such doting care to thrive. She lives a mostly healthy life that will likely be longer than most working farm hens—of a comparable span to those in our own urbane and demanding flock. Her nourishment is scratched up, nibbled, and hunted—or provided as a morsel forgotten by a careless human. She is free to wander as she pleases, at night roosting concealed among low palm fronds or perched upon the warm roof of a tour bus. To lay, Cluck will simply select a quiet spot and produce two or three smallish eggs that will hatch into smallish chicks. Together they will spend their days loitering at an outdoor mall or darting among hotel lounge chairs on a never-ending prowl for tiny bugs and dropped ice cream cones. They pose for occasional photos but are wise to keep a low profile, lest they end up on the menu.

It seems that everywhere you look in Kauai, a chicken is lurking just on the edge of paradise.

Opposite: Cluck warily escorts her brood through a Hawaiian garden, always with an eye out for danger—or food.

A SOFT-BOILED IDEA

"Maybe our hens could live like Cluck," we mused that night. It seemed like a reasonable notion at 2 a.m. A Kauai-style, semi-feral flock would require less time and care, save us lots of money on feed, and provide the old girls ample opportunity for healthy exercise.

But such jetlagged musings were not so reasonable in the light of day. We knew that our Portland, Oregon, climate was relatively temperate but very cold compared to this island, with seasonal light and temperature changes that make food periodically scarce and stress unsheltered hens to the breaking point. Foraging alone, even in summer, would never provide the concentrated nutrition domestic chickens need to lay a good number of sizable eggs. And that's not to mention the murderers' row of local coyotes, hawks, and raccoons awaiting their opportunity to enjoy an easy chicken dinner.

We might not be able to simulate the relative easy life of a tropical climate with few predators for our own chickens, but we intuited that there was much we could learn from plucky Cluck and her chicks. Unlocking her secrets could help us improve how we feed our flock, better protect them from predators, heal our injured and sick, and work within their enigmatic pecking order to foster harmony.

HARD-BOILED TRUTHS

Shortly after our return to the mainland, we learned of impending changes to federal regulations governing the availability of poultry medications administered via feed and water. Meant to discourage the overuse of antibiotics as a growth

stimulator on commercial farms, the Veterinary Feed Directive (VFD) would also limit access to many remedies on which home chicken keepers rely as treatment for seriously ill chickens. Effective and inexpensive treatments for coccidiosis, various respiratory and gastrointestinal infections, and certain parasites would soon require an expensive trip to the vet, chicken in tow, to obtain and fill a prescription. We knew that these changes were sure to ruffle a few feathers but suspected that the impacts would be more far-reaching than most folks realized.

Chickens on large commercial farms are usually maintained in as nearly aseptic conditions as possible to limit their exposure to disease and protect their fragile immune systems. They spend their short lives consuming a finely calibrated diet and may receive a variety of pharmaceuticals to treat disease. In stark contrast, Cluck and her friends on the island seemingly radiate good health despite, and because of, their decidedly more chaotic milieu. Their vitality starts with diverse and robust genes and is supported by a spacious and stimulating environment. Compelled to forage their food from a variety of plant and animal sources, they are constantly exposed to an astonishing diversity of immune-supporting—as well as challenging—microbes.

The life of an urban backyard chicken resembles both extremes in some respects, and the effectiveness of our care is similarly middling. For instance, typical coop

A Hawaiian mama hen relaxes in the safety of the underbrush with her chicks.

and run maintenance practices reduce odors and can be effective at setting the "bad bugs" back, but they may never enable a favorable microbial equilibrium to become established. Over time, a proliferation of pathogens can remain in the soil for years, silently sickening and sapping the vitality of multiple generations of chickens. Similarly, providing organic feed supports sustainable agriculture and may limit our own exposure to toxins, but for our hens, without the stimulation and nutritional benefits of foraged food, it does little to enhance their vitality or that of our gardens. So, rather than resist the new regulations, we embraced them as a call for more widespread adoption of a comprehensive, prevention-based approach to caring for our birds.

SUNNYSIDE-UP CONCLUSIONS

As soon as we arrived home from Hawaii, our daughters made a beeline to the coop to reunite with their hens, stopping only to collect for them a welcome basket of plump worms on a bed of lush weeds from our organic garden. Our girls squealed with delight as their frenzied hens devoured the offerings, and we sighed contentedly, struck by the intimacy of our family's relationship with our flock.

Admittedly, it's not always such an idyllic picture. Our daughters have gone out to visit the ladies expecting a cheerful greeting from a favorite hen, only to find her cold and motionless in the corner—a death that might have been avoided with a few dietary tweaks during a heat spell. And neighbors have anonymously reported us for a loud rooster we'd kept a few weeks longer than planned.

We learn as much from our hens on these sad occasions as we learn when things are going smoothly; both situations provide feedback we need to tweak our approach to new challenges while trying to remain true to the values that first attracted us to keeping chickens. As we learn and evolve, so, too, do our customers at the Urban Farm Store, challenging us to aid their quest for effective and sustainable solutions for their small-scale chicken-keeping needs. What we all want is a fresh way of keeping chickens, one that honors both wild Cluck and domesticated Checkers.

Reflecting on our experience with the chickens of Kauai and what set them apart from our flock at home, it became clear to us that our own long list of seemingly individual problems with our chickens had a few common roots. Excluding pests and predators and keeping our hens where we wanted them were reasonable and essential goals, but the traditional methods of achieving them with fortified, immobile coops could potentially set the stage for serious behavioral, health, and dietary problems. It was as though our chickens were being kept in only two dimensions when they truly need a third to thrive.

We now realize that we, and you, can improve an overall home flock setup by making it more dynamic, so that our chickens can safely forage for food and scratch around, but when and where we want them to. Such a flock functions as

Hens that live in harmony with their environment glow with vitality.

part of a larger whole, as key performers in the dance of life that is an organic garden, where they are the grazers and recyclers of waste, harvesters and conservers of solar energy, and miraculous producers of fresh and nourishing ultra-local food.

This book, with a renewed focus on a prevention-based, organic approach to care, is for any chicken keeper, novice to advanced, who seeks better results from a more integrated approach to chicken husbandry. Some of you will try to adopt all of our recommendations. Others will read about a range of options that can help you reap as many benefits as possible without taking the big plunge and embracing the free-ranging, organic philosophy all at once. We provide plenty of tips for those who choose to keep their hens in a stationary coop.

Distilled in these pages are our twenty-five years of combined personal and professional experience keeping chickens and a map for our shared road ahead. We invite you to join us on the journey.

BIRDS *of a* FEATHER

THE NEED FOR BREEDS

Knowing full well the popularity of chicken keeping in Portland, we were startled to read a 2013 US Department of Agriculture survey that conservatively estimated that more than 600,000 nonfarm American homes hosted flocks of chickens in urban and suburban contexts. The survey also revealed that although less than 1 percent of households had chickens at the time, about 4 percent of folks without chickens planned to start raising them within the next five years. That's a lot of new chicken keepers, and a lot of chickens!

Nearly 20 billion chickens inhabit our planet, making them the most populous livestock by a factor of ten over their nearest competition, cattle. The majority of these birds are raised for meat, and about 75 percent of those are produced using "intensive" methods, a euphemism for "factory farmed."

It's sobering to consider that this agricultural juggernaut, so vital to our global food supply, relies heavily on a very shallow gene pool consisting almost exclusively of a single type of chicken, the Cornish Cross. This hybrid merges a short, but thickly muscled, British breed with a longer-boned, more upright Asian type for the singular purpose of efficiently converting cheap feed sources to meat. Within massive, environmentally controlled buildings, these birds park themselves at a feeding station, start eating, and don't stop until they are ready for market (or succumb to maladies caused by their rapid growth), going from egg to grocery store in less than two months.

The welfare and genetic-diversity issues are similar in the egg industry. In the United States, in the month of October 2016, 300 million individual hens cranked out a record 7.51 billion eggs. The vast majority of these birds are variants of a single breed, the Leghorn, a heavy layer of white eggs that originated in the Mediterranean region.

As intensive agriculture has advanced globally, many traditional, locally adapted landrace breeds (domesticated breeds that have developed distinct features over time in response to their environment and by being isolated from other populations of the same species, rather than through formal breeding) have been abandoned in favor of modern breeds tailored for close confinement

A flock with a variety of breeds is beautiful and practical.

In a typical factory farm, crowding chickens with low breed diversity in high-density conditions and a highly artificial environment practically invites health problems.

Left: These chickens will be marketed as cage-free. Old MacDonald would not approve.

Chinese Silkies share a backyard coop with a hybrid cousin.

Above right:
Modern breeding and agricultural systems are displacing traditional chicken farmers, who depend on regionally adapted and diverse flocks.

and intensive feeding. When these landrace breeds vanish, with them go the genes that contain the many unique traits for which they were selected, along with untold millions of others with additional attributes that will be increasingly valued in a future marked by climactic shifts and resource scarcity. The loss is also cultural in the sense that each breed is a reflection of the values, environmental settings, and even aesthetic tastes of the humans who developed it.

In a hopeful sign, a few major producers of commercial chicken breeds are beginning to reverse the trend of genetic simplification by incorporating more genetic diversity and developing breeds for alternative farming systems such as pasturing. Hy-Line International, an old-school heavyweight breeder and hatchery of Leghorn hybrids, now offers a range of breeds that produce white, brown, or tinted eggs and are better equipped for living longer and thriving in more humane and sustainable alternative production systems.

Although we applaud these modest advances by the big guys, it's clearer than ever to us that the remaining dual-purpose generalists (breeds that provide both meat and eggs) and other unique, regional, and odd breeds are the sacred inheritance of the backyard and small-flock chicken keeper. Our support of these breeds is their greatest hope for survival. This need not mean, however, that we

THE IMPORTANCE *of* QUALITY BREEDING

Our store sources chicks from several hatcheries that offer reasonable breeding quality at prices that our customers are willing to pay. Feed stores typically offer this quality level or lower—an important factor to consider when seeking birds to establish or expand your own flock. Although average-quality hens can make serviceable pets and layers, skillful breeding reveals the best features of a breed, making hens more likely to be terrific layers, tame pets, or a good fit for whatever traits have drawn you to them. Recent research has revealed that, beyond aesthetic and productivity considerations, disease resistance and even life expectancy are heritable and will be heavily influenced by the quality and diversity of the breeding, nutrition, and the conditions in which parent hens are kept. Happily, these factors favor reputable breeders who take the time to develop good bloodlines.

At a minimum, we recommend that you inquire about the particular hatchery that supplies chicks if you purchase them from a feed store (and ask about vaccines while you're at it). Do a little research on the hatchery. How many breeds do they offer? Where are they located? Do they keep their own breeding stock—and, if not, what can they tell you about the farms that supply them with hatching eggs?

Chicks of truly rare breeds (or high-quality lines of more common breeds) regularly exceed U.S. $200 each, with some individuals of exceptional quality fetching prices in the five figures. Although it's not necessary, or even advisable, for beginners to spend this much, we encourage veteran chicken keepers to experience for themselves the true potential of well-bred chickens and to support small breeders' efforts to maintain their breeds and preserve unique poultry genetics for future generations.

If you do opt for chicks of higher quality or rare breeds from craft producers, be aware that these will be available only unsexed, as straight-run, meaning that you're likely to get both males and females. Have a plan ready in advance for what to do with the males if they are not allowed in your area.

The Ayam Cemani, a beautiful and rare breed, is completely black, from its feathers to its internal organs and bones.

must limit ourselves exclusively to keeping so-called heritage breeds, because they can also be combined and recombined as new breeds to improve vitality and increase diversity.

Of the more than 400 breeds of chickens that inhabit the world, only a small fraction is available through domestic breeders and hatcheries in the United States. By some estimates, only about 70 chicken breeds (plus color variants) have been established here during the 500-year importation history of poultry in North America, and most of those are slowly vanishing, displaced by a tidal wave of modern hybrids bred specifically to live in intensive commercial farming systems.

BEST BREEDS FOR THE BACKYARD

From the easy-to-appreciate, fluffy, blond Buff Orpington, to the slightly bizarre, stilt-legged Malay and buzzard-like Naked Neck, we find something to appreciate about all chicken breeds. Our favorites for backyard chicken keeping are known for their utility, appearance, temperament, novelty, or a combination of those traits.

Much has been written about commonly available breeds, so we don't devote much space to those here, other than naming our top picks. Instead, we decided it would be much more interesting, not to mention critically important for preservation, to highlight a few unusual and endangered breeds that are well suited for small flocks and that deserve more attention.

With so many interesting breeds, we found it difficult to limit ourselves. If we had more pages and time, we could wax poetic about the intricate patterns of the Laced Wyandotte (including the ultra-rare Blue Laced Red colorway!), the handsome and intrepid Brahma that was bred to thrive in the demanding environment of the Himalayas, and fine Mediterranean examples such as the prolific white-egg-laying, large-bodied Catalana hen from Spain that rarely gets broody and is very tolerant of heat. To conserve space, our description of each breed's main attributes is brief. Let careful consideration of your own unique location and needs guide your choice of breeds.

Though each breed has at least one particular standout trait, we nominate the Plymouth Rock and Ameraucana as all-stars that can do it all—from laying eggs and turning compost to surviving the rough handling of a toddler. We've always recommended that new chicken keepers start with some common breeds like these and move on to more exotic options after they have a little more experience. That said, if you must have a rare and beautiful specimen like a Lavender Orpington or Blue Laced Red Wyandotte and you find one for sale, by all means, nab it!

ALL STAR ★

PLYMOUTH ROCK

Outstanding backyard hen

PROFILE

- 6 to 8 pounds
- Attractive, commonly available, and genetically robust
- Early and highly productive layer of large, light brown eggs
- Potential year-round layer given suitable conditions
- Quick to mature with early graduation to coop

- Easily tamed and adaptable to family life
- Excellent forager and garden helper
- Hardy in hot and cold, robust, and long-lived
- Somewhat vanilla personalities
- Hearty appetites must be balanced by activity to avoid obesity

Barred Plymouth Rocks are attractive and productive dual-purpose chickens.

Rocks, as they are usually known, are a relatively old American breed that was likely derived from a series of crosses between the similar French Dominique and one or more Asian breeds (possibly Java, Dark Cochin, or Brahma) in the mid-1800s. Their broad genetic base has proven highly advantageous, enabling breeders and hatcheries to preserve much of their utility and vigor to the present day. Only two colors are commonly available—barred (stripes of black and white, almost checkered) and white—but several more color variants persist and are worth seeking out. Generally speaking, the barred strain is favored as a dual-purpose breed, and the white is raised primarily for meat (though they are also good layers), as is often the case with white breeds, because of the cleaner-looking carcass produced after plucking. Rocks were crossed with Cornish chickens to create modern meat hybrids.

Before World War II, hatcheries produced specialty strains of this breed that were collectively the most common egg layers and meat-producing chickens on U.S. farms. Perhaps this was because of their very early age of first lay and quick rate of growth. When they are fed sufficient protein, you can expect to see their first pale brown eggs a staggering three to four weeks before most other breeds. Early physical maturation makes possible an early and welcome exit from the brooder to the coop, typically a week sooner than other dual-purpose breeds and up to two weeks earlier than smaller-bodied, white-egg-laying breeds. As a meat bird, they are considered to be early finishers for a heritage breed, but not when compared to modern hybrids.

These truly hardy birds usually take cold weather in stride, often managing to lay straight through moderate winters (if not molting). Remarkably, we've observed that they also tolerate heat fairly well if given shade, water, and moist earth. This versatility earns them a top position among other breeds for ease of care. We also marvel at their ability to thrive in urban confinement and on free-range farms; along with the Rhode Island, Rocks are a common sight in both environments. They love to dig and scratch, making them ideal members of hard-working flocks in mobile coops and pens.

Despite their work ethic, their personality is fairly calm, they are easy to hold and stroke, and they can be readily tamed as pets. Their large bodies also ground any aspirations of flight and slow their walk, so they are more likely to stay where you put them and are easy to catch when they don't. Barred Rocks are known to get along well with other hens and other well-behaved pets, and they are always a favorite of children, perhaps because of their resemblance to zebras.

Barred Rocks are an ideal component of a mixed flock, with plumage that's visually complementary with that of other breeds. Their eggs possess both a distinct color (pale brown, almost pink) and shape (slightly pointy) that enables easy differentiation in the nest.

Barred Rock chicks are semi-autosexing, which means they can be sexed with fair accuracy. The mostly black fuzz on a young female is marked by a faint white dot on her head. This becomes less distinct over time, but males may be identified by more white on their wing tips and a generally lighter appearance.

AMERAUCANA

The only commonly available breed that rivals Rocks for our affection

PROFILE

- 4 to 5 pounds
- Blue-tinted eggs are unique and appealing
- Prolific year-round layer in most climates
- Relatively hardy and disease resistant with few congenital problems
- Genetically robust
- Good temperament with potential to tame as a pet
- Newfound popularity may lead to careless breeding in the near future, undermining genetic quality
- May be difficult to find a specimen of the true breed (hybrid Easter Eggers are common and passed off as Ameraucanas)
- Not meaty enough to be truly dual-purpose
- Better flyer than larger breeds; may need wing feathers trimmed

The bearded Ameraucana is a prolific layer of blue-tinted eggs.

Ameraucanas are easily confused with Easter Eggers and are sometimes sold by this name, but, technically, Ameraucanas are true-breeding layers of eggs in a range of blues, while Easter Eggers are similar-looking hybrids that lay slightly more green-blue-tinted eggs. Adding to the confusion, the Ameraucana Breeders Club is confident that the Ameraucana is not simply an improved version of the rumpless (without a tailbone and often without tail feathers), similarly blue-egg-laying Araucana from South America, but is in fact an entirely separate breed that was standardized in North America in the late 1970s. It now seems likely that most of the birds sold as Ameracauna or Araucana are neither, and are indeed some sort of hybrid Easter Egger. For the purposes of most home chicken keepers, they are essentially the same bird, although we would select a properly labeled Ameraucana over an Easter Egger any day, and twice on Sunday.

Our Ameraucanas are, without fail, our best layers each year, and customers and friends usually report similar results, which is somewhat surprising, considering most were purchased for a bit of egg color novelty. This exceptional laying ability is probably in part a result of being a recently developed and skillfully bred line, meaning that the breed has not gradually lost vigor over many generations of line breeding with limited stock, like some of the heritage breeds.

Some males may achieve admirable size, but Ameraucanas are not considered top-quality meat producers. Though they are not quite as hardy as some of the big European chickens, we've observed Ameraucana to be exceptionally healthy and long-lived in our temperate backyard setting.

CLIPPING WINGS

If you have a heavy breed hen, you may not need to clip her wings, or you might just clip one to make flights more awkward and embarrass your hen into staying put. For the rest of the breeds, such as Ameraucanas, you will likely need to trim some flight feathers to keep them in unroofed runs or in your yard while they forage.

Our definition of a heavy breed is any large bird that you can actually catch. It usually lays brown eggs. If you can't catch it, and particularly if it lays white eggs, it's probably not a heavy breed. If it flies over the fence and is never seen again, just shake your head and mutter knowingly, "Those darn flighty Mediterranean breeds. Should have clipped both wings," as you gaze in the direction that you imagine it flew.

Wing clipping is a simple process if you have a helper, a good pair of sharp scissors, and a couple of headlamps. Start out at night, while the hens are sleeping; they will remain in a torpor and are much easier to handle in the dark. After your partner plucks a hen from her roost and gently (but firmly) holds her with wings tucked against her body, take hold of a wing. Extend it to reveal the five or six longest feathers that look like they would make good quill pens for writing. Looking from the feather tips toward where they attach at the meaty part of the wing, locate where the shaft thickens and usually changes color. Do not cut below this area. Using sharp, full-sized scissors, cut off all of these feathers about halfway down from the tips, so they are roughly even with the remaining small feathers on this side. There should be no blood, but if there is, stop and reposition the cut a bit further out from the main wing (use styptic powder for the rare times bleeding does not stop almost immediately on its own). Repeat on the other wing for increased effectiveness.

Clipping our crested Polish hen's flight feathers is painless and is often the only practical method for keeping our free-ranging chickens in the yard.

RHODE ISLAND

The iconic chicken of the backyard and homestead

PROFILE

- Prolific layer of large brown eggs
- Few health problems and a strong constitution, even as a chick
- Terrific forager and efficient consumer of feed
- Infrequently broody
- Adaptable to confinement or free ranging
- Can be a bit high strung, but with a little extra attention, she'll be your best garden buddy
- Males raised for meat mature in 14 to 18 weeks, up to 8 pounds

This famous American heritage breed actually does hail from Rhode Island, where it can trace its ancestry to a single black-breasted red Malay cock that is now preserved for posterity in the Smithsonian Institution in Washington, DC. Appropriate for an American breed, this chicken's pedigree is a motley blend of immigrant breeds from East and West, with Java, Shanghai (now known as the Brahma), and the prolific Brown Leghorn from the Mediterranean rounding out the mix. Rhode Island Red roosters contribute the paternal genes for various types of sexlink hybrids, so called because males and females can be distinguished at hatch by their feather color or markings.

The deep mahogany plumage characteristic of the breed is not the only color available. There's a treasure trove of genetic potential and vigor embodied by the lesser-known Rhode Island White. Produced by making a fresh but similar mix of Asian and Mediterranean breeds around 1900, the white chickens are reputed to surpass the reds in laying and meat production. This breed was identified by The Livestock Conservancy as being threatened, with less than 3000 birds counted in their 2015 poultry census. The original and darker Rhode Island Reds are becoming rare as the breed is "improved" to meet industry needs.

True Rhode Island Reds are a deep mahogany color.

BARNEVELDER

Consistent layers of large, dark brown eggs

PROFILE

- Prolific producer of up to 200 eggs per year
- Steady temperament, docile, and easily handled—great candidate for taming as pet if handled often while young
- Well adapted to confinement or free range
- Fairly cold hardy
- Good forager
- Slow to mature
- Males raised for meat mature in 20 weeks to 8 pounds

Above: Barnevelder plumage is gorgeous, and their deep brown eggs nearly match their feathers in color intensity.

This large, 200-year-old Dutch breed is still admired for her dark chocolate-brown eggs and intricate feather patterns. The breed is a sound dual-purpose (meat and egg) chicken with its robust blend of European and Asian genetics. Roosters are particularly handsome with plenty of meat on them, though their full weight of nearly 8 pounds is not reached until the relatively old age of 20 weeks.

SUSSEX

A True British Classic

PROFILE

- Excellent layer of cream-to-tan-colored eggs
- Friendly and easily tamed as pet
- Great forager
- Adaptable to confinement or free range
- Very cold hardy
- Sensitive to extreme heat and humidity
- Males are meaty, at 6 to 8 pounds

Sussex checks all the boxes—a hardy, attractive, heritage breed that is notable for meat and egg production. Originating in the cool North Atlantic region, the breed can be sensitive to very hot, humid weather, and for this reason the Sussex is not recommended for the U.S. Deep South or subtropical areas. They are unusual as a surviving pure landrace U.K. breed—they lack Asian genes common in other breeds.

The gorgeous speckled brown feather pattern is most common in the United States, but several breeders now offer the more unusual color variants such as the Light Sussex. We encourage you to support breeders' efforts by selecting these variants if you see them. This endangered heritage breed needs help to survive.

Departed but not forgotten—our beloved Speckled Sussex, Sweet Baby Chicken

A couple of breeds are new to us or otherwise noteworthy.

Our Cream Legbar, Agnes, struts her stuff.

CREAM LEGBAR

Cream Legbars have been a particular favorite of ours since we first ordered them as hatching eggs to start a breeding flock a few years ago. Although that project was derailed when the store's roof collapsed in a 2014 snowstorm, we've enjoyed keeping one of the survivors and admire her good looks and other attributes. She gives us lovely, pale blue eggs and has tireless foraging habits. Best of all, this European breed reliably exhibits the ultra-rare autosexing trait that enables us to sort males from females at a glance at any age. That's a big plus for individuals and communities that want to achieve independence from hatcheries and reach new levels of sustainability.

Hard to identify by appearance alone, all Whiting True Blue hens lay light blue eggs.

WHITING TRUE BLUE

Though somewhat similar in appearance to Ameraucanas and Araucanas, this modern breed is unrelated to them, and their eggs lack the green tint of Easter Egger eggs. They can be difficult to identify by appearance, because the individual chickens can have virtually any feather color. We have recently begun keeping Whitings at home for evaluation, but if the claims are accurate, Whiting True Blue hold great promise as another reliable layer of blue eggs.

Our customers searching for pets and child-friendly breeds have usually done some research and are drawn to breeds identified as being especially docile. In this case, "docile" may refer to ease of human handling, but it may also indicate that they are easily dominated by other breeds in a mixed flock. Perhaps a more specific term for a good pet chicken would be "tame," which better captures our desire to handle and interact socially with chickens as we do with other domestic pets. It's been our experience, however, that individual chickens of any breed can be tamed by frequently and gently handling them as chicks and teens as well as hand-feeding the adults, but it's true that some are easier than others. Along with the best general breeds for backyard chicken keepers, we recommend several others for those who are more interested in companionship than egg production.

COCHIN

Plump and Pleasing Balls of Feathers

PROFILE

- Can be kept in a small yard
- Low flyer that stays within fence
- Very broody, good hen for hatching eggs
- Not a good forager, comparatively gentle on gardens

- Produces few eggs
- Robust and cold hardy
- Prone to overheating
- Hardy appetite and low activity can lead to obesity

Cochins are astonishing: large, absurdly round, and comically covered in feathers on almost every surface, including heavy feathering on the tops of their feet and their legs. The Cochins we've known are dense, warm mounds of soft feathers that genuinely seem to enjoy interacting with humans, almost like a dog or cat. We suspect that this apparent affection is only partially because of breeding, and more likely a byproduct of their slow, earthbound bodies that make it impossible for Cochins to evade the grasp of their keepers, and in this way they become accustomed to handling. They never protest being picked up and hugged, and their plush feathering positively screams, "Pet me!" These low-flying birds are disinterested in flight when food is around, so they'll stay put in almost any pen. In fact, they are often too heavy to reach high perches, so be sure to place theirs at 2 feet or lower.

The arrival of Chinese-bred Cochins in Britain in the nineteenth century ignited the poultry craze that still reverberates today. The early crosses are the origins of many of the well-known heritage breeds. It appears that Cochins were originally

The adorable Cochin is a great choice for those who want more pet, less livestock.

bred with ease of handling, or possibly even petlike tameness, in mind, as they are supremely well suited for it. Nevertheless, most hens will produce a few eggs each week in nice weather. We recommend pairing them with a couple of hens of other docile breeds that lay consistently if you also want eggs.

Their feathered feet can become encrusted with mud, making them susceptible to infection and foot diseases, but covering pens in rainy climates and good litter management can help keep problems at bay. You can soak affected hens and trim their feathers if necessary to remove mud and debris. In addition, in hot weather, it's advisable to clip the fuzz under the tail and around the vent to keep the area clean and less prone to fly strike.

Cochin chicks are known to pile up as they sleep, suffocating each other in extreme cases. This is often caused by competition for a small spot of warmth in the brooder and can be prevented by warming a larger region more evenly—but avoid heating their food and water.

Like any quality collectable, this breed is coveted for its many desirable color variations and unique forms. For starters, their amazing plumage can comprise virtually every pattern and color—you could almost match it to your décor. Many colors are also available in bantam size in frizzle, a quirky, kinked feather type that also helpfully eliminates any remaining possibility of flight.

Cochins International maintains a fantastic website that offers a remarkably complete guide to caring for these unusual chickens and their special needs.

ORPINGTON ·

The Golden Retriever of Chickens

PROFILE

- 7 to 8 pounds
- Moderate layer of light brown eggs, may lay sporadically through winter
- Prone to broodiness

- Matures quickly
- Well adapted to small spaces
- Good breed for colder climates
- Gets along well with flockmates

· ·

Orpies come in a variety of colors, but by far the most popular is the buff, with lovely yellow-gold feathers.

The Buff Orpington is at the top of most lists of chickens to keep as pets, though we've found them unremarkable in this regard. Before you send us an angry e-mail in rebuttal, keep reading. We have seen many marvelous pet Orpies—but it's a matter of which came first, the reputation or the reality. Most other breeds become just as tame when in the care of the type of person who likes to hug chickens. That said, we will concede that they are better layers than Cochins, though most Orpingtons are being bred as pets these days, and scant attention seems focused on preserving or enhancing laying abilities. Nonetheless, we appreciate the Buff Orpington for her color, good looks, easygoing personality, and winter hardiness.

The breed was developed in England by crossing a Minorca with a Black Plymouth Rock. At one time, Orpies were champion layers, capable of cranking out more than 300 eggs a year, but the breed has since declined shockingly in productivity as a result of careless breeding.

If you can find a quality breeder, Orpies can be spectacularly attractive—almost as round and statuesque as a Cochin without the sometimes problematic feathered feet. Seek out a color variant, as these tend to be more carefully bred—and satisfying. The Lavender Orpington is the most sought-after novelty color these days, but the breed is also available in black, white, jubilee (speckled), spangled (mottled), and probably a few more.

We've had better luck finding good layers among their cousins from down under, the Australorp (the Australian Orpington breed), a cross of the Orpington with the Rhode Island Red and several others.

MIXED-FLOCK MYSTIQUE

We are often asked by customers whether it's okay to mix chicks of different breeds, to which we respond, "Why would you not?" Although breeders and serious livestock folks have good reasons for keeping pure flocks, diversity has many advantages for the backyard chicken keeper.

Try combining breeds by selecting those with complementary habits. For example, keep hens that will lay moderately year-round with others that are more seasonally abundant to ensure a steady supply of eggs throughout the seasons. Or combine strong layers with more petlike breeds. Having a diversity of breeds enables you to learn firsthand about behavior differences and see how well each breed is suited to your environment and lifestyle to inform future selections and recommendations.

The distinct egg colors from a variety of breeds also make for a stunning egg basket, and the rainbow of egg colors are also useful for tracking the laying tendencies and output of the individuals in your flock. If one of your motivations for keeping chickens is the delicious eggs they produce, you may be surprised to learn that eggshell color has no direct effect on flavor (although fans of particular breeds such as the Maran may disagree), and the color of the hen has nothing to do with the color of her eggs. Egg flavor is one variable you can safely cross off your list when deciding which breeds to seek out.

SEXLINK *and* AUTOSEXING: WHAT'S THE DIFFERENCE?

Urban flocks are typically prohibited from including roosters, and even larger rural flocks need only a few, if any, to sustain them. Accordingly, female chicks that have been pre-sorted from their male counterparts are desired by practically everyone who keeps chickens to produce eggs. Sounds straightforward, but the difficulty arises because male and female chicks within most breeds look identical until they have matured and it's too late to sort them. To offer their customers all-female chicks of most breeds, hatcheries must rely on trained personnel to isolate them from males based on minute differences in their sexual anatomy, visible only briefly after they emerge from their shells.

Some breeds, however, helpfully exhibit some degree of sexual dimorphism (physical differences between sexes) in feather color or pattern that makes sorting by early feathers alone possible, though not highly accurate. Hatcheries produce chicks with more reliably distinct markings by crossing breeds in specific ways to create sexlink hybrids.

The best-known examples of sexlinks are the gold sexlink, black sexlink, and red sexlink (sometimes sold under less racy-sounding names, such as Gold Comet). None of these chickens are true breeds, however—they are first-generation (F1) hybrids, and mating them will produce unpredictable results, both in terms of feather sexing and other important traits such as egg production.

Very early feather differences have been stabilized in some breeds to the extent that they are autosexing, meaning that this trait will reliably appear in each generation they are bred. Such breeds include the Cream Legbar and Bielefelder.

Black sexlink hens are a cross between a Rhode Island Red rooster and a Barred Rock hen.

·· »

THE GROWING BROOD

Our first chicks were a somewhat impulsive purchase. Sure, we did plenty of reading, formed opinions about which breeds seemed best, and usually had some sense of how many adult birds would fit in the coop we were planning. But when the fateful day arrived, it took only one peek into that first bin overflowing with fuzzy, colorful baby chick goodness to kiss those plans goodbye. We wanted them all! This is how many a beginner finds herself sheepishly returning to her car, balancing a huge sack of feed in one arm and a small box brimming with twice as many chicks as originally planned in the other.

Even the most experienced chicken keepers find their springtime visits to the brooders at our store practically irresistible. If we listen closely, we can hear their rationalizing above the din of tiny cheeps: "There's always room for a few more," or "We've got all of the gear already—might as well use it!" And my personal favorite, "I've been looking for this breed for years."

Can you resist?

WHAT ROOSTERS, OFFICER?

We've long wanted to establish a self-perpetuating flock of heritage, dual-purpose birds at home to produce both eggs and meat. The problem has been the rules, common to most cities, prohibiting keeping of the roosters that we would require for breeding.

We've considered trying to game the system a bit by keeping a few young roosters just long enough to breed. When they're young, they don't crow quite as enthusiastically as they do when they get older, so in theory they wouldn't bother the neighbors enough to report us (again). Our idea is to start with two or three male chicks of a desired breed late in the growing season—late September or early October in our area. They'll be fully feathered and will have gained enough size to survive outside by January but will not reach sexual maturity (and get noisy) until around March, when our hens, kept year after year, begin laying in earnest again. At that point, hopes high, we'll start pulling a few eggs from the nest each week and popping them in our incubator or under a broody hen. After we have confirmed the fertility of six to twelve eggs, we'll harvest or relocate (if our daughters protest too much or for further breeding if they prove to be exceptional) the now loud and illegal roosters before the neighbors decide they've had enough of nature's alarm cock (pun intended).

By nature's design, about half of the chicks we hatch will turn out to be roosters. Of these, we will eat or relocate all but the finest two, which will be ready to breed by September, starting the cycle over again. As needed to replace lost hens, plan for succession, or expand the perennial hen flock, we can keep one or more of the female chicks and the rest can be sold or given away.

If the plan works, we'll enjoy a lovely flock of dual-purpose, heritage chickens genuinely breeding like a real flock. If not, it may be time to switch to plan B: find an urban rooster stud service!

Roosters are handsome, but noisy.

Take a moment to slow down and consider your motivation for keeping hens in the first place. And remember, whether you're establishing a new flock or adding a few fresh hens to boost egg production, overpopulation is an invitation for trouble of various sorts. After many years, Robert can scan a backyard, size up the coop, tug on his beard thoughtfully, and intuitively arrive at a number: "You can have up to six chickens in here, but you seem like a busy family and I'd suggest four. If the kids help and your neighbor over there will watch them when you're gone and chip in on the feed bill for some eggs, you can handle six." Making this judgment for yourself requires evaluation of available time, feeding resources, and the amount of both yard and coop space you can dedicate to the new additions. Feeding and time constraints are mostly cost-driven factors, but most of us can use a little help sizing up our spaces.

Coop size itself is seldom the most limiting factor, considering that chickens require little room for sleeping and laying. Each hen needs only 8 to 12 inches of space on a roosting bar, and a 4-by-4-by-4-foot coop can include at least two bars if they are offset and at different levels—enough to house eight to ten hens and three nest boxes. This density comes at a maintenance cost, however, because poop will accumulate at an alarming rate from so many birds, necessitating near-perfect ventilation and frequent cleanings to maintain low ammonia levels and a healthful condition within. If pressed to provide a formula, we'd say that each standard-sized hen requires a minimum of 8 cubic feet of coop space, but you'd be wise to make it 12. An often overlooked, yet crucially important, limitation on flock size is the run and foraging area.

MINIMUM COOP REQUIREMENTS BY FLOCK SIZE

Coop size in feet (W×D×L)	Maximum capacity	Preferred capacity
4×4×4	8 standard / 10 bantam	5 standard / 7 bantam
4×4×6	12 standard / 16 bantam	8 standard / 12 bantam
shed, 8×8×8	24 standard / 32 bantam	16 standard / 24 bantam

A flock of sixteen hens may fit comfortably within a modest coop shed but will wreak havoc on a landscaped yard if they are allowed to run amok there daily. We figure that a 4-by-8-foot stationary run is the bare minimum for two or three hens, and they'd be far healthier and easier to care for in a much larger area, ideally 25 to 50 square feet per bird. If you don't have that much land to dedicate to a run, consider a mobile coop that can spread its impact over a large area while maintaining a compact footprint.

PICKING UP CHICKS

GUIDELINES FOR A HEALTHY BROODER

On Kauai, chicks must grow fast and move even faster to keep up with their busy and highly mobile mamas. With abundant year-round forage, warm daytime temperatures, and the protection of their mother's downy belly nearby, these chicks live in ideal conditions for their growth and behavioral development—that is, if they manage to avoid being eaten by an egret, bullfrog, dog, or cat. Compared to the Hawaiian chicks, hatchery-bred chicks are hothouse flowers, and we are their gardeners. But chicks are more resilient than they seem, and as long as you remain vigilant and follow some basic guidelines, your chicks will thrive.

Chicks are easy to care for indoors in a brooder but can be dusty. Don't brood your chicks in areas near where food is prepared, in sleeping quarters, or in other places where the brooder could pose a health or cleaning issue. A basement, mudroom, spare bathroom, and cool garage are all good bets. Be sure to locate the brooder near an outlet to avoid foot-tangling extension cords when possible. Regardless of where you put the brooder, you'll appreciate having a fairly large spot that's easily accessible.

Don't think that you can brood young chicks outdoors, even in the summertime. The primary advantage of brooding them in your home is the steady, moderate temperature and the fact that you'll be in close proximity—a big advantage for maintenance and monitoring for potential problems. The exception is using a spare coop for a transitional brooder, which offers welcome relief from the unruly antics of teenaged chicks, typically kept inside until they reach eight weeks or so.

AVOIDING HAZARDS

Consider hazards within your home when placing your brooder, and keep a few easily overlooked hazards in mind.

- Dogs and cats, even when supervised, may suddenly attack or accidentally injure chicks. Although they may look cute together, it's better to keep them separated.

- Children are ideally suited to be assistant chick caretakers, but they must first learn safe handling skills and safety procedures.

- A fall can be fatal while you're holding a delicate chick; tripping over a brooder box on the floor can also be hazardous to both you and your chicks.

Although most chicks are fine in brooders made from a large plastic tote, and dog crates are also a popular choice, something as spacious as a large cardboard box offers your little ones the room they need to regulate their own body temperatures, stretch their legs and wings, and practice natural behaviors such as foraging and dust bathing. A large box will also help you maintain ample separation between a heat source and the chicks' food and water, avoiding the warm and damp conditions that incubate pathogens. A large brooder box also gives your chicks time to grow larger before you move them outside, which is particularly beneficial if you will be integrating them with adults.

Our hands-down favorite jumbo brooders are galvanized metal stock tanks, often sold at feed or farm stores and used to water animals. In the urban landscape, you've likely seen these deep metal tubs used as garden planters—or as bathtubs for cigar-chomping cowboys in old westerns. They are an ideal size and are extremely durable and not flammable, an important consideration when you're using heat lamps to keep your chicks warm. In a pinch, and with a little imagination, rabbit cages and other pet gear can also be repurposed, but the fact remains that bigger is better when it comes to brooders.

In the mood to brood! A good brooder setup, like this one in a galvanized metal stock tank, offers a sturdy enclosure, clean water, absorbent litter, and a carefully placed source of regulated warmth.

LITTER MATES

Screened wood shavings used for horse stalls or small animal cages are commonly available litter materials that work well. We avoid cedar, however, because it can be toxic to young chicks—look for pine, poplar, or aspen shavings. Purchase enough shavings to fill the bottom of the brooder 2 to 3 inches deep.

Our favorite medium to scatter on the brooder floor is coir, a highly absorbent material made from coconut husks. We originally found it marketed for reptiles, but it makes a superior, highly absorbent litter as well. Look for the coarse, chunky grade sold for horticultural use and reptile bedding. Coir is sold in compressed bricks (to reduce shipping cost) in two weights: a dry, larger chunk kind often labeled as husk chips, and a fine, shredded fluff. The chunk type is far easier to break apart when dry than the fine grade, which must be wetted prior to using. We recommend the chunk type, because moistening the shredded coir bricks defeats the coir's utility for chicken litter. The chunky stuff holds plenty of moisture and is a prized addition to the garden as well—before or after serving its time in the coop. As with wood shavings, coir should be spread to a depth of 2 to 3 inches.

As quaint as it looks, straw should never be used to brood chicks. It usually performs poorly and often harbors mites. This also applies to the interiors of all but the largest coops. Straw is great stuff for outdoor runs, however, where it helps to control odor and combat mud.

OPTIMIZING BROODING ENVIRONMENTS WITH LIVING LITTER

Having raised a few dozen of our own chicks to maturity and after tending to hundreds more at the store, we confidently believed that we were doing it about as well as possible and thought there was little room for improvement. But as we reflected on the lessons we learned from the Hawaiian chickens and recent changes in medication regulations, we began to suspect that there was more we could do during this crucial period of growth and development, from hatch to about twelve weeks, to lay a foundation for lifelong chicken health and resilience. As it turns out, the missing piece of the puzzle would come not from our chicken-keeping experience, but from an entirely unexpected source: our backyard beehives.

Honeybees gather plant resins and use them to produce propolis, a sticky substance known to possess antibacterial properties. We read with great interest recent findings that revealed how propolis conveys social immunity to the entire colony, enabling individual bees to invest less energy in their immune systems. We knew intuitively that a similar effect must protect wild chickens, an insight that could provide a new approach to how chicks are brooded at home.

Coconuts are a major cash crop in tropical countries, particularly India and Indonesia, where this versatile fruit is used to produce both food and fiber products. Millions of pounds of coconuts are processed each year, creating mountains of coir fiber behind processing plants.

To explore our hypothesis, we perused scientific research, skimmed a few old farming manuals, and attempted to catch up with the efforts of the folks who share their approaches and debate topics like this online. We concluded that wild chickens also gain social immunity from the microbiome (the microorganisms in a particular environment, including the body) of their flock and its environment and that some of these benefits could be simulated by chickens keepers, similar to the way farmers traditionally raised several generations of healthy chicks on the same litter.

We're not suggesting that you save soiled litter from your brooders. It's far easier and more predictable to create living litter using materials that help fresh litter function more like a naturally aged material. Based on our research, we now recharge our home brooder with a few ounces of biochar (a form of charcoal that neutralizes toxins in feed and improves litter conditions) and several ounces of activated beneficial microorganisms (such as EM, a proprietary blend of

LIVING LITTER CHICK CARE

Age	Temperature (warm zone)	Area per chick*	Food, hydration, treats	Bedding, environment**	Health & wellness
Hatch to day 7	90°F	2 sq. ft.	Unmedicated chick starter food with 20% protein, pinch of fine grit, sprinkle bokashi over food	2–3 in. depth, non-aromatic wood shavings, coir, or blend of coir and shavings Add biochar, bokashi, perches, swings, other features	Observe and treat for pasting-up
Weeks 2–3	80°F	2–3 sq. ft.	Aloe and EM/BM in water		Monitor for deformities, feather picking, wounds, parasites heat stress, weight loss, foul-smelling or bloody poop
Weeks 4–5	75°F		Continue above, plus scatter variety of tidbits over litter—coarsely ground grains, whole garden greens and weeds with roots and soil bits, mealworms (earthworms, may contain parasites dangerous to chicks)		
Weeks 6–7	65–75°F (or into outdoor coop or brooder***)	3 sq. ft.			
Weeks 8–9		3–4 sq. ft.			
Weeks 10–11					
Weeks 12+	Outdoor coop	4 sq. ft. or more	Transition to 18–20% protein developer feed, or blend ⅔ starter plus ⅓ layer rations		Continue, plus watch for predators

*More is better. We often exceed this density at our store without harm, however, by monitoring and using other management protocols, including cages that enable the poop to drop through the cage floor into a tray below.

**Assumes use of nipple waterer for chicks that allows for our litter management approach. Standard litter management will require more frequent changes.

***Small numbers of teenage chicks will need moderately warm night temperatures or a heat lamp to keep cozy outside in a coop or brooder, but five or more can huddle in cool nights and can go out as early as five weeks old in suitable conditions.

BIOCHAR BASICS

Biochar is food-grade charcoal produced by burning wood or other carbon sources in a reduced oxygen environment. It's not permitted as a commercial feed ingredient because it can mask the presence of contaminants, but you can add it to food and litter to protect both from many toxins. We sometimes sprinkle biochar on chick and adult hen food, but we use it mostly as a booster in our coir-based litter. Biochar is a powerful odor and ammonia absorber, and we are convinced that it is helpful for maintaining healthy litter conditions, particularly when litter gets deep and ages. The hens actually seem to seek out and eat the biochar in the litter, perhaps using it like grit, scratching and fluffing the litter in the process in a way that helps maintain healthful conditions.

beneficial microorganisms) incorporated into coconut fiber, changing it only when necessary, if at all. Though it's difficult to confirm that we have seeded our chicks' guts with healthful fauna, we know that we're cleaning the brooder less often!

The table on the previous page summarizes our current (and evolving) blend of standard, new, and experimental techniques for living litter chick care.

WATERING YOUR BROOD

The traditional function of litter in a brooder is to absorb moisture and keep odors in check. Although a variety of materials are effective, all seem destined to find their way into the drinking water as the chicks kick up litter while practicing their foraging and dust-bathing skills. More than an annoyance, the resulting stagnant and poopy water, coupled with the warmth of the brooder, has the potential to encourage growth of pathogens that cause illness and sap vitality.

Although the risk of hazardous water contamination is lessened by raising the water above the litter, it is not eliminated unless it is coupled with the use of drinking nipples, which are designed to release small amounts of water on demand from a reservoir. This system relies on gravity and the instinct of chickens to peck at shiny and colorful things.

Until recently, if you wanted to use nipple waterers with chicks, you had to make your own device using individual nipples purchased for custom applications. Small nipple founts are now available to purchase and are the perfect size for brooders. To ensure hydration for our chicks, we start the first day with a regular fount plus a nipple fount in our brooders, and then we remove the regular fount the next day. Nipple waterers are not only very efficient, but because they are installed overhead and rarely, if ever, leak, they have essentially eliminated the problem of poop commingling with drinking water and remaining there to fester. Drinking nipples are the secret weapons that enable farmers and chicken keepers to let their litter accumulate without forcing their livestock to drink warm poop tea. We strongly recommend their use; otherwise, the usual admonishments to raise the water above the litter and clean it often apply to the living litter method. Whether you use nipples or an open waterer, any unpleasant odors from the brooder are a sign that something is amiss and the brooder needs cleaning.

PROVIDING LIGHT

Chicks need light to stimulate their pituitary glands for proper growth and development. Natural, indirect light provided by a nearby window is good, but avoid direct sunlight that could overheat them. Windowless basements, which are ideal locations in other respects, are too dark unless you add artificial lighting from broad-spectrum lights that mimic natural sunlight. Regular light bulbs will work in a pinch, but we don't recommend using them as a source of light, or warmth, in your brooder. Light sources should be put on a timer that will turn them on and off to simulate day and night.

KEEPING YOUR PEEPS WARM

Chicks also need supplemental warmth to maintain their body temperatures and metabolize their food properly. This is naturally provided by the downy feathers of mother hens, but you can provide an artificial source with an infrared heat bulb in a heavy-duty lamp base positioned above the brooder. These lamps and bulbs are often sold in feed stores and hardware stores, especially in the spring. Hang the lamp by its cord, or use the supplied clamp to attach it over the brooder. Set it at a good distance to produce a warm zone in part, but not all, of the brooder at the appropriate temperature for the chicks' age. For warm locations such as inside a heated home, use a 100-watt infrared bulb; for cool areas such as a basement or garage, use a 250-watt infrared bulb. Because these heat sources need to remain on for twenty-four hours a day, use red infrared heat bulbs to avoid interfering with light cycles; the red bulbs may also help reduce aggressive behaviors.

If your brooder is large enough, the heat source distance and brooder temperature need not be as exact as they would be if using a small brooder, because the chicks will have room to enter and exit the warm spot to regulate their body temperatures, similar to how chicks use a mother hen's warm belly. Try a combination of thermometer and chick behavior to determine setting and distance. If the chicks avoid the warm spot completely, raise the heat source to lower the temperature; if they huddle together under the warm spot and seldom leave, lower the heat source to provide more warmth. A large brooder and a warm zone also enable you to place food and water in a cooler area to keep it fresher.

Infrared bulbs emit lots of heat, but true infrared heat sources are a unique technology that warms bodies, not the air in between the bulb and the bodies.

"HONEY, DOES MY CHICK BROODING PLUMAGE
MAKE ME LOOK FAT?"

The bulbs are superior for warming chicks, but if you put your palm in front of one to judge the heat output, you might not feel as much warmth as you'd expect.

Always use caution with high-wattage bulbs—especially the 250-watt size, which is hot enough to ignite litter and cardboard and melt plastic. Suspend it above the brooder and avoid clamping the lamp to flimsy sides, and never remove the bulb guard that is sold with most quality lamps. Always use a lamp rated for high wattage bulbs in an appropriate receptacle.

A good infrared lamp is more expensive than a regular brooder bulb and lamp, but it will be long-lasting and durable, so you can use it on your third, fourth, and fifth expansions of the flock. It will still be serviceable the next time you brood and can be used safely and efficiently to warm adult hens in their coop.

Another warm alternative that's discussed at online chicken-keeping forums is the electric hen, which provides a little heated shelter for chicks. You can purchase ready-made electric hens or make your own. Several online sites offer instructions on how to make electric hens by placing an electric heating pad over an arched wire screen or other heat-resistant support that's sturdy enough and high enough for chicks to gather underneath. Use a name-brand, moist-heat–rated pad without the auto-shutoff feature, and make sure it's in good condition, with no tears in the plastic pad cover. It's said to be safe for use in this application, even when sprinkled with a protective layer of litter on top—as always, use caution when repurposing an item originally intended for a different use.

HEN HABITATS

SUCCESSFUL SHELTERS

Around 6:45 in the evening, the already languid pace of life on Kauai, where this journey started, slows to a complete standstill. You head to the lanai for a nip of rum and sunset appreciation, and peer over the railing. If you're lucky, you may spot green sea turtles hauling themselves out of the surf for their moonlight snooze. Raise your gaze to the palm trees and you are likely to see and hear the raucous arrival of the mynas as they gregariously congregate between fronds high over-head. Look in the lower, bushier trees and there, about 15 feet above the ground, are Kauai's feral hens settling in small groups on the bare branches, with one or two roosters on guard below.

This is how domestic chickens would shelter if they lived in a tropical climate with no tree-climbing predators. Alas, most of us do not live in a climate that remains moderately warm year-round, nor is our environment free of climbing marauders. Our mainland hens are mostly pleasingly plump egg layers and not lean flying machines capable of reaching the safety of high branches. We do know of at least one fellow who manages to maintain a population of smaller bantam breeds of chickens in his orchard in a milder part of Oregon, but it's hardly what we think of as chicken keeping: he seldom finds any eggs and periodically experiences heavy losses to predation. For the rest of us who intend to keep our chickens alive for several years, the only alternative is some form of confinement.

SITING THE HABITAT TO AVOID RUFFLING (HUMAN) FEATHERS

When asked to picture a chicken coop, most folks imagine a small shack clad in weathered boards, perhaps surrounded by a large fenced pen over a patch of bare ground. To us, however, that describes two separate things: a coop and a run.

Coops are where your chickens come home to roost, to take shelter from preda-tors and harsh weather, and to sleep. This is also where hens lay their eggs—if you

are lucky and they don't choose a hiding spot elsewhere in your yard. Coops should be sturdily constructed, with solid walls, a roof, and a base; insulated if needed; and with some form of ventilation. Urban coops range in size from a large kitchen appliance to a walk-in shed.

Runs are the fenced areas that surround the coop, where your chickens spend most of their waking hours eating, dust bathing, scratching around in the soil, and squabbling . . . er . . . socializing. They may be partially or completely roofed. Runs vary widely in size, from tiny rabbit-hutch–sized cages (sometimes integrated with a small coop), to entire backyards enclosed by a tall wood fence.

The true art of chicken keeping involves striking a balance between the benefits that protection provides and the difficulties that arise from keeping hens confined in close quarters for extended periods of time. Consider our first chickens, Rosy and Roxy, who lived the good life. Each morning, a human arrived to let them out of their coop and tend to their food and water before releasing them to roam their lush, fenced yard all day. Just after nightfall, a human was dispatched to secure them in their coop, safe from predators and out of the chilly night. Their days were filled with fresh air, plenty of exercise, and unlimited nutritious forage to supplement their main diet of quality feed. Whether hustling for worms as we dug in the veggie patch or lounging with us on the patio, our girls were charming garden companions. And they laid delicious eggs with perky, deep orange

At dusk, the chickens of Kauai prepare to return to their roosts.

Together, coops and runs protect and contain urban chickens.

yolks that tasted like creamy sunshine. This simple approach seemed ideal, and we were bitten by the chicken-keeping bug, and bitten hard: we enthusiastically expanded the flock from two to eight birds in our first summer together as a married couple.

The trouble started quietly with the squishing of cold poop between barefoot toes, followed by the unpleasant discovery of a cache of rotten eggs the girls had hidden from us under a hedge. Then, alarmed by the wilting of a prized blueberry shrub, we parted the limp foliage to expose an unrepentant Rosy digging her muddy claws in the soft soil, shredding roots and snacking on bugs while expanding a moist hole where she cooled her belly on hot afternoons.

Soon our little pack of hooligans was uprooting whole plants, and when our lawn browned in the summer, several managed to get over the fence to munch our neighbor's much greener lawn. Most depressing of all, we soon realized that much of the backyard and patio had become so peppered with droppings that we were avoiding it and spending more and more time on the chicken-free front side of the house.

The final blow to our free-range utopia came in the form of two deadly daytime attacks: the first was an inside job by a dog that we were dog-sitting that should have been avoided, and the second was an entirely unexpected assault by an off-leash hound that somehow got under or over the large wood fence that surrounded our yard.

Stubbornly resisting any suggestion of abandoning free ranging, we convinced ourselves that fencing half the yard for the hens and keeping the other half, including the patio, for ourselves would be a fine compromise. We recruited renowned store employee Pete (a genuine farmer!) and the hole digging, post setting, and fence tightening commenced immediately. All told, we installed more than 100 feet of fence and gates to enclose a spacious and protective run for our flock that comprised almost 2000 square feet. We somehow managed to stay within 300 percent of the original budget—and within a month of the estimated completion date!

Not wanting to block off half of the yard with an expensive and forbidding solid fence, we opted to use rolls of relatively inexpensive wire fencing stretched over a light wooden frame. Despite Pete's considerable strength and knowledge of fencin', the fence bulged and the tops sagged flaccidly wherever it spanned more than a few feet. On a farm, you could pass this off as rustic charm, but in urbane Portland, it just looked shoddy. Nevertheless, on the glorious day we installed the

last of the gate latches and put the hens inside, we held a glass of Oregon pinot noir aloft for a hearty toast: "To protecting our remaining chickens and sitting on poop-free chairs, and to that ugly fence that will make it all possible."

The winter after we installed the fence, new challenges emerged that would dog us into the summer: a spike in the local rat population, an uptick in the number of nights we forgot to close the coop door, and the loss of more plants within the enclosure from intensified digging and pecking. Worse yet, the sagging top of the fence made a perfect launching platform for our gals to reach the neighbor's lush lawn, and their raids actually increased. It was clear that our neighbor's patience was wearing thin, probably not helped by our acquisition of a hive full of bees that liked to drink from, and drown in, his hot tub.

We purchased and installed a $300 automatic door for the coop to cover for our forgetfulness, but that did little to disguise the fact that there were deeper problems. Underlying the mounting malfunctions was a truth neither of us could yet admit: the physical separation created by the

fence had weakened our connection with the flock, and with it went some of our enthusiasm. We still loved and enjoyed having our hens, but for much of the day they were simply out of sight and out of mind.

It was clear that we were visiting the girls less often, and with less handling they started to lose some of their tameness, further weakening our bond. After our first child was born, there were long stretches in that soggy and harsh winter that we visited the hens only when absolutely needed to refill a feeder or their muddy water fount. Our hens were in danger of becoming livestock rather than beloved egg-laying pets.

Top: Fenced out of our own yard

Bottom: Fencing provides safety, but it separates hens from the true sources of vitality in the garden beyond—as well as our companionship.

ROOM TO ROAM
· ≫

Had we not moved to a new home later that year, our flock may have faced an entirely different danger endemic to the long-term backyard chicken keeper. Chickens (and other animals) housed in small yards for many years may begin to exhibit otherwise unexplainable declines in productivity or suffer chronic disease problems capable of spanning generations of new hens. This is usually caused by

A large mobile coop and fenced run offer the best of both worlds: free range and safe shelter.

long-lived and persistent disease-causing pathogens in their environment that have become entrenched, infecting and reinfecting hens as they are recovering. In a dark twist, some individual birds will respond to medication, giving the appearance of successful treatment, even as whole yards silently become contaminated with potentially antibiotic-resistant organisms that eventually make chicken keeping nearly impossible. If the situation is dire enough, the toxic environment will need to be abandoned. For one unfortunate friend of ours, the somber prescription was just such a clean slate: new chickens lodged in a fresh coop located as far as possible from the infected housing. He burned the old one.

In response to the danger of antibiotic resistance at both the farm and backyard levels, the U.S. Food and Drug Administration enacted the 2015 Veterinary Feed Directive (VFD), a move designed to transition the availability of penicillin and other antibiotics for food-producing animals from over the counter to by prescription only by 2017. As we became acquainted with its provisions and considered their impact on our customers, we began to think more deeply about organic methods of thwarting organisms that can cause diseases in chickens. We concluded that reducing pathogen loads and breaking the cycle of reinfection were the environmental keys.

While delivering feed to our customers, we began to compare the variety of coops and methods of chicken keeping we encountered, judging them both by

OUTLAW CHICKENS

We are obliged to remind you to research local laws, homeowner's association rules, and your rental contract (if you do not own your property) regarding chicken keeping before you decide to start or expand your flock. Most urban and suburban areas allow some number of chickens to reside within city limits but regulate the number of birds you can have, coop locations, and the level of care. However, some cities, like Portland, will issue you a permit to keep additional chickens after your yard and coop have passed an inspection and you can prove that you have met certain other requirements. Virtually all cities prohibit roosters because of noise, but you may be able to keep them as well for a short time before they reach maturity.

We know it's hard to believe, but there are some American towns and cities where it is actually illegal to keep chickens. We would never (ahem!) want to imply that these laws are in any way elitist or misguided, so we'll just say that we respectfully disagree. Our message to our oppressed brothers and sisters is this: Stay strong! The day is coming when your hens can come out of the closet and lay in the open air of freedom!

Chicken-discriminatory laws have been easily toppled by a handful of activists in such places as Athens, Georgia; Tampa, Florida; and New Haven, Connecticut. Having reviewed several of these cases, we can reveal the secrets to the pro-chicken activists' success: 1) Chicken keepers fighting city hall makes for great human-interest news, and 2)

if the law is really oppressive or old, so much the better as far as ease of having a new one drawn up is concerned. One example of a wacky law comes from New Hampshire, where state law once required that chicks be purchased in a minimum quantity of a dozen, effectively preventing potential chicken keepers from starting small flocks or sending folks on clandestine missions to the borders of Maine or Massachusetts to get their chick fix. The law was changed easily enough when a neighbor of a state representative simply asked him for help. Not only did he convince the legislature to make the change, but he got three chicks himself!

We cannot openly advocate going rogue, but keep in mind that chicken laws are typically enforced on a complaint-based basis. Happy neighbors and a discreet, well-tended coop will virtually guarantee that no one is going to come looking into your chicken operations. We're not suggesting that you disguise your hens with bunny ears and whiskers or dye their eggs bright colors—it's usually enough to keep your chicken coop clean and out of sight from the street to avoid having people ask questions. Do involve your neighbors in your love of chicken keeping by stressing the positives. Assure them that your chickens won't smell bad and that you plan to keep only a few hens and no loud roosters. Ask if they have any other questions or concerns. If absolutely necessary, you can make limited promises of eggs—but be sure to let them know that having enough to go around may require you to get even more chickens!

their outward appearance and by their owners' testimonies whenever possible. It became obvious to us that the healthiest, longest lived, and most productive chickens all had plenty of space to roam, either freely in large runs or confined in mobile coops. Like our wild friend Cluck, these hens had ample access to fresh forage and naturally dispersed much of their own waste before it could accumulate in concentrations conducive to disease.

Despite their many advantages for both chickens and their keepers, mobile coops have never been as popular as the familiar stationary coop-and-run setup, which the vast majority of home chicken keepers opt to use. So why doesn't every chicken keeper use a mobile coop? Our personal chicken-keeping saga (continued on page 68) provides the answer. Although practical lifestyle considerations and aesthetics often win out in urban and suburban settings, we urge you to allow your flocks to free range or use mobile coops. If this is simply not possible, some complementary strategies for feeding, probiotic supplementation, and poop management will go a long way toward tipping the balance away from disease and in favor of populations of beneficial microbes, improving the health of your hens from the outside in.

"I'm an egg donor."

KEEPING PEACE IN THE FLOCK

Mixing chickens can be difficult, both among chicks of different ages in the brooder and when newly brooded young hens are moved into yards with an existing flock. Chicks of different ages can be mixed in a brooder if their ages are not more than about two weeks apart. After that, perhaps only very large brooders with lots of hiding places for younger chicks will work, but it's better to set up a separate brooder and integrate them later using one or more approaches. Mixing different breeds of similar-age chicks is seldom a problem.

When chicks are ready to move outside, where established older hens await them, your juveniles *will* be bullied. The pecking order is a very real and natural thing (not just a human expression), and what we may see as brutality is a somewhat unavoidable process of deciding who's in charge and how the rest of the order will shake out below that top hen on the totem pole. As difficult as it may be for you to watch, your role is to stay on the sidelines and observe the action, intervening only if you notice blood or a serious injury. There are a few things you can do to help out, however.

Who's in charge here?

+ If you have a second coop, appropriate day pen, or even a large dog crate, set it up within a larger coop or outside the coop sharing a side of the coop. Put the new chicks in the second coop so the birds on both sides can become acquainted. We're not huge fans of this approach, because it can be difficult to maintain the separate flocks, plus it simply delays the inevitable day when they will be combined. Nonetheless, this is the approach most commonly used and some folks swear by it.

+ Instead of separating (or after using the separated approach for a while), we've had good luck by providing lots of fresh hiding places along with the new birds. Added terrain such as a jumble of thick branches will slow the attack by established hens and give the new girls a place to take a break. Adding a few perches to the run will also help, providing high resting places away from attackers. Getting cornered is very dangerous for a young hen, and you must make adjustments within the coop and run as needed if hens often seem to be getting trapped in a particular spot.

+ Distractions and boredom busters are our big keys to integration. Try adding a whole bale or big flakes of fresh straw to the run, sprinkled with treats such as cracked corn, and toss some fresh weeds on top for a finishing touch. A large, compressed treat block is also compelling and should buy some time for the new gals to become a normal part of the scene.

+ It can be very effective to set up the new girls in a mobile pen or coop with one or two of the less aggressive older hens, or otherwise use the mobile unit to break up the established cliques so the social order is not so clear and enforceable.

+ Never add a single hen to a group. She will have no ally and will be more severely treated. If you must, try pairing a new hen with a docile hen in separate quarters for a while before reintroducing both after a few weeks.

These suggestions also apply when integrating older hens to an existing flock, whether mature adults or started adults (pullets). From a planning perspective, you will likely need a secondary coop or pen and a redundant set of feeding and watering gear for the newcomers for several days to weeks. Separating new hens also enables you to inspect them for health issues before they mix with your existing flock. This precaution is the best way to protect your flock from disease and parasites.

BOREDOM BUSTERS

Chickens are intelligent and curious, and, like most other animals, your hens need an interactive environment that provides opportunities for stimulation. Zoos and other captive animal care facilities use the term "environmental enrichment" for the many methods they use to provide mental and physical activities and challenges for animals in their care. We call it "chicken play."

You can offer your chickens boredom-busting play options in a variety of ways that can complement almost any style and setup. Free-ranging and mobile chickens usually get enough stimulation from foraging, but the more confined your hens, the greater their need for play. You can provide play opportunities regularly, or add them as needed to create diversions, such as when you are cleaning the coop or introducing new hens. The basic goal is to provide a stimulating and complex environment to prevent boredom and to encourage social participation and bonding.

TOUGH CHICKS

It's not uncommon for us to get a frantic call from a person raising chicks who is concerned about a bully bird in the brooder. We gently ask whether there has been any injury to the menaced chick or if it's just disturbing for the caller to watch. If there is no blood, we assure the caller that it is normal for one chick to dominate the others and suggest the addition of a few boredom busters (interesting foods, toys, or other stimulating attractions) to redirect her behavior. If she has seriously injured a brooder mate, we instruct the caller to tend to the wounded chick first and then move the perpetrator into her own quarters for a few days (she'll need her own warmth, food, and so on). She can then be returned to the main brooder in a couple days to see if her attitude has improved.

If taking a time out doesn't help pacify a bossy chick, you'll need to take other measures. Larger chicks and adult chickens can be fitted with tiny chicken blinders, or peepers, devices that resemble spectacles and painlessly attach to the nostrils on the beak. They work by blocking part of a hen's vision to prevent her from targeting a blow with her beak, while allowing enough vision to the sides to enable her to continue eating and drinking normally. If a chick is too small for peepers, you may consider a trick that our coworker Jeremy recommends: He suggests gently binding the aggressive chick's legs together with string, a small bandage, or surgical tape (being careful not to bind them so tightly that the chick cannot walk). This works either because her attack is slowed or because she is embarrassed—we're not quite sure which.

Attaching chicken peepers to the beaks of aggressive hens keeps them from focusing on and pecking others.

FOOD PLAY

Feeding chickens novel foods offers the most common—and useful—method of play. Chickens' lives revolve around food, so they are inclined to participate enthusiastically. Try lengthening feeding times: Instead of tossing them diced apples for a treat, hang a whole one from a string. Or go the other way—cut that apple into tiny bits and scatter it over a large area. Hang a whole cabbage just above their heads, and your hens will spend hours pecking at it. Food can also be used to train chickens to participate in other types of play.

One of our favorite foods to encourage play is a compressed treat block. Available in several sizes and sold under various names, a treat block is a mixture of grains, seeds, and nuts glued together with molasses into a solid mass that stimulates repeated, persistent pecking. They are available in various sizes and are extremely compelling and attractive to hens (and lots of other animals). A treat block will instantly become the focus of hours of pecking, which is especially useful because the most dominant hens tend to get most involved eating and defending it, distracting them from harassing other members of the flock. Blocks are an excellent tool for creating a diversion when introducing new hens, as well as a quick fix for redirecting antisocial behavior long enough to break bad habits such as toe and feather pecking. It's important to realize, however, that this is junk food: it is very high in energy and low in protein, so laying and other health aspects will be affected if hens eat it for more than a few days at a time. Blocks are also a favorite with rats and mice, so it's practically required that you put them away at night.

Our chickens like to peck at the bright colors of a toy glockenspiel—and it keeps them entertained for hours.

SENSORY PLAY

Enrichment using light, sound, touch, and smell also seem to be entertaining and stimulating for chickens and vital for their well-being. For a challenging olfactory workout, try hiding favorite treats (preferably those with strong odors) in a place where your hens will need to sniff them out. For visual stimulation, place a shatter-proof mirror in the run or on the other side of a fence—your hens will be very curious about it and usually find it good fun to look at the new friends they see there, but watch for signs of territorial distress—they might peck at it and hurt themselves.

Adding a bale of straw to a chicken run is a perfect stimulator for touch. Hens use their clawed feet for essential scratching activity, while their beaks probe and sort the material. This is our hens' very favorite activity—and the minimum amount of enrichment we'd recommend providing.

If your flock is prone to pecking, give them something they're encouraged to peck. Hannah mounted a colorful metal toy glockenspiel on the side of the coop and is teaching the hens to use it by luring them with treats, like she trains our dog.

Make a nature playground with rocks, jumbled logs, and branches. Add ladders, overhead perches, and small platforms for climbing and sanctuary for smaller birds. Make or purchase a swing, or just string-up a branch with some rope.

Your hens will "dig it" if you give them sand or dirt piles. Pen your hens over the compost bin, or bring wagonloads of it to them to explore.

COGNITIVE PLAY

Puzzle feeders are a good example of cognitive enrichment. They provide mental stimulation for hens and lots of cheap entertainment for their human audience. Parrot owners have been working on this for longer than we have, so we recommend drawing inspiration from their websites and forums, where you can learn about how they challenge their highly intelligent pets with foraging blocks, climbing nets, and toy challenges. Be aware that chickens are often not as agile as other birds and may get stuck or otherwise be unable to use pet bird gear. It's a good idea to allow them to play with these toys only when supervised until you are convinced of their safety.

It's also possible to train your chickens with the same positive-reinforcement techniques popularly used with dogs. A simple example is training your hens to come running when you call them. Just say something like, "Heeerre chickeee, chickeee," dribble a little cracked corn by your feet to the waiting hens, and repeat. After a few minutes, or at most a few days, your hens will learn to associate the sound of those words with the treat and will be trained to return with the fidelity of a Labrador retriever. Using the same basic technique, you can train your hens to perform a variety of more complex behaviors and even, in some cases, extinguish undesirable behaviors by consistently ignoring them while reinforcing the desired behavior with treats and sounds.

SOCIAL PLAY

Chickens are social animals that naturally live in small groups with a hierarchy we call the pecking order. By keeping more than one hen, you are providing essential social enrichment, and, though we don't suggest you invite other hens over for playdates (there would be aggression, plus this practically invites disease problems), it can be fun to spice things up by pairing hens with rabbits, for example; they are similar in size and are generally peaceful together, though we recommend that you supervise. Cats are also interested in hens and are perfectly safe around them (exceptions include smaller bantam breeds, youngsters that cats can overpower, and exceptionally aggressive hunting cats). It's probably a good idea to leave other species of birds off the guest list whenever possible, for biosecurity reasons.

Roxie prepares to dismount from her homemade chicken swing.

Opposite: Mixing it up by including a rabbit or two can help keep chickens entertained.

TWO COOPS *are* BETTER THAN ONE

Our first coop and run seemed huge when we first erected them, but it took only a few months to realize they could not contain our burgeoning chicken fascination. We struggled to enlarge the area, but we knew that, ultimately, we'd have to start over in a new yard to scale up properly for our hens' needs. To spare others this fate, we generally recommend that newbies build from the outset for twice the chicken housing capacity they anticipate.

For those faced with the daunting task of enlarging a coop, we say this: Don't bother, because you are far better off adding a second coop and run—or, better yet, a mobile coop. A redundant setup of some sort will provide unique flexibility that one large coop will never match. Our auxiliary coops are invaluable as transitional stopovers for younger birds bound for the main flock, refuges for injured and bullied hens, and quarantine areas for new birds from other flocks.

HAZARDS OF THE TRADITIONAL
URBAN COOP, AND HOW TO FIX THEM

The chicken-keeping arrangement we encounter most often consists of two or three hens housed in an improvised coop surrounded by a small fenced run. These hens commonly spend their entire lives within a small patch of backyard real estate, occasional garden sojourns notwithstanding. Most of the hens seem energetic and well cared for, but some appear to resent this tight confinement, passing their days languidly scratching at their half-clean litter, pecking disinterestedly at stale food, and sipping murky water from a dangling, rusty fount.

In time, this regular manner of keeping chickens in the city inevitably produces a range of difficulties, stressing the well-functioning setups and swamping the rest. We've seen many chicken keepers become disenchanted with their pursuit, unable to keep pace with the effort needed to battle an endless array of tiring chores and recurring problems. It can be hard to perceive, but what we are experiencing is not a series of unrelated problems, but the slow-motion disintegration of an over-simplified system.

Although we offer a somewhat radical prescription to improve this situation, much can be done to make progress on the most vexing issues with a few simple changes. Rather than list and attempt to solve every housing-related problem we routinely encounter, we offer broader solutions that aim to shore up the entire system and prevent problems before they start.

MANAGING POOP

Our wild friend Cluck's poop produces no particular hazards, because she spreads her waste over a large area. Some chicken keepers can emulate her environment by keeping hens in low densities and keeping them moving, but for those with less space and time, confinement is the only option. They must use litter to collect, absorb, dilute, and transport their hens' poop.

This is an important issue, especially for those whose hens stay within small coops or yards. Realize that chickens are eating and pooping machines, with each hen producing 4 to 6 ounces of moist poop daily, and this waste exits directly into their environment. Naturally, there's nothing fundamentally wrong about this, and these rich droppings are a marvelous source of fertility for the organic garden. But within the cramped environment of a coop and small run, droppings can accumulate quickly. If not addressed, they will begin to stink and attract flies almost immediately, and they may pose a serious health hazard. So, without further ado, we present to you the straight scoop about the poop in your coop.

Along with undigested bits of feed and waste products from normal metabolism, a hen's poop contains harmful pathogens she's trying to shed. In cramped spaces or with high population densities, waste can rapidly accumulate, forcing your hens into close contact with this material and increasing the likelihood that they will reingest (eat) it, thwarting the healing process. A main way we unknowingly expose our hens to this risk is by feeding and providing water to our hens within the coop, in the mistaken belief that hens require access to food and water while laying or after turning in for the night. Rest assured that keeping food and water elevated and sheltered outside the coop will keep it close at hand (beak?) and greatly reduce the risk of cross-contamination. Elevating the food and water is also a basic strategy to help avoid contaminating what your hen eats and drinks with fecal matter.

And that brings us to litter, an essential ingredient for civilized chicken keeping that is quite literally required by urban chicken keepers.

MANAGING LITTER

We sell mountains of wood shavings and straw each week, which rank a close second behind layer feed in total sales. Of course, there's a very good reason for this: the function of litter is to absorb moisture and keep odors in check by neutralizing nitrogen before it becomes ammonia gas. We share our customers' concerns about the sustainability of litter and hope to popularize viable alternatives such as hemp and coconut fiber. Whichever litter type you use, we encourage you to compost it to use as an amendment to your garden soil.

We usually advise chicken keepers just starting out to keep the litter management routine uncomplicated by using the deep litter method, which uses clean, dry, and pest-free litter material such as wood shavings. The technique is simplicity itself: Begin with a couple of inches of fresh shavings on the interior coop floor, and maintain it weekly by adding a few fresh handfuls of new material. Restart the process with fresh shavings when the litter on the coop floor becomes too deep or odors become noticeable.

This approach works well because the airspace within the shavings enables the litter to resist molds and mildews while absorbing moisture without becoming matted down. To avoid trouble, keep lower-performing litters, such as straw, leaves, and woodchips in the outdoor run, where moisture factors matter less and where the mites they can harbor will not have the same opportunities to infest your flock.

You can take it to the next level and boost the effectiveness of this or any litter management system with biochar, bentonite clay, and our homemade antimicrobial herbal oil spray. When used together, these will help you maintain a litter condition that's unfavorable to microbial growth and toxic to pathogens. In addition, adding beneficial microorganisms, such as those used in bokashi composting, reduces odors and harmful ammonia gas while injecting trillions of good microbes that dominate your coop's doo-doo dojo.

MUCKING OUT THE COOP

Mucking out the coop is sometimes avoided as a nasty chore, but we choose to focus instead on the results: a bounty of nutrients, microbes, and organic material to benefit the garden. To make your harvest easier, you can try several tools and coop modifications.

BLENDING DIET *and* ENVIRONMENT WITH BOKASHI

The two aspects of our flocks' health and vitality most within our control are their diet and their environment. In our early days instructing chicken keepers, we presented diet and environment as two entirely distinct factors, but as we observed our flock over the years, we've come to realize that the two are invariably linked. Hens compulsively hunt through yards, coops, runs, and litter searching for food (even those fed a complete ration) and are constantly encountering their feces, and that of other animals, in the same places in which they are foraging. Considered from this perspective, our hens' diet and environment ought to be viewed as a single set of environmental inputs that we need to manage in concert for optimal health and productivity.

Managing the environment amounts to managing our hens' droppings using litter. In our experience, chicken keepers maintain the litter within their coops and runs in either a fluffy and dry (aerobic) condition with frequent cleanings, or in a more dense and moist (anaerobic) state that relies on regular but infrequent additions of litter to offset the accumulation of droppings. These methods both aim to stay ahead of pathogens and odors with additions of fresh litter, varying chiefly by frequency and intensity of cleanings. By our estimation, both methods seem to consume an equal amount of the chicken keeper's time and effort in total, and both manage odors, but it's unclear how they differ in effectiveness at producing a healthy environment.

Regardless of how often you clean your flock's habitat, keep in mind that it is already teaming with microbes, including some that cause disease and others that play vital roles both internally and externally in maintaining chicken health. In simplistic terms, the conditions you create in your coop and run will either favor the good guys that foster health or tip the scales to the bad guys that cause disease and drain your flocks' long-term vitality and productivity.

To stack the deck in favor of the desirable microbes, we spike our hens' environment with regular additions of the beneficial microbes used in bokashi composting, such as EM—Effective Microorganisms, a proprietary blend of beneficial microorganisms. If you're not familiar with the term, *bokashi* is a Japanese word that refers to an anaerobic fermentation technique that uses blends of complementary microbes to pickle (acidify) organic waste products. Its chief advantages over traditional composting are that it does not rely on huge piles of material to heat up and it works best in the worst, muckiest conditions.

We should be clear that if your coop litter is very airy, true fermentation will not be possible. Nonetheless, the diversity of microbes present in these probiotic blends will ensure that at least some of them will find conditions to their liking. Where good microbes thrive, undesirable microbes, and diseases and odors they cause, cannot.

It is possible to achieve something more akin to true bokashi results in the low-oxygen conditions of deep, moist, and infrequently changed litter. Start by "activating" a gallon of the microbial blend to awaken from dormancy by following the manufacturer's directions. Blend it with 2 to 3 cubic feet of shavings (and/or coconut fiber) and pack it tightly in the floor of the coop. Replenish this weekly with a thinner, additional layer of the same material and pack it tightly. Your hens will fluff it up again, but you can pack it down from time to time.

We also use bokashi in our kitchen to capture the nutrient value of our food waste that's unsuitable to feed directly to our hens. We collect and bokashi food scraps in a bucket in the kitchen, and then bury the fermented results in our garden after a few weeks. We love that bokashi produces neither odor nor the loss of nutrients that occurs with regular composting—it's free fertility to turbocharge our garden and our hens. After a few weeks, the buried fermented material mellows and becomes incorporated within the soil. The contribution can be significant: last year our bokashi contributed an astounding 720 pounds of nutrient-rich material to our food, ornamental, and container gardens!

One person's chicken poop is another person's garden compost.

Tips for managing muck

+ Muck rakes are aptly named, long-handled tools that feature plastic or metal tines angled to make scooping litter much easier. Use one to scoop moist litter, straw, chunky coir, or other material; or use it to sift through dry material as you would when cleaning a cat's litter box. Wheeled muck buckets are also sold at feed and farm stores; they make transporting the poop to the compost pile much easier and cleaner.

+ If you are into harvesting chicken poop to use as garden fertilizer, or you just want to catch most of it where it accumulates the fastest, install a droppings pit, board, tray, or hammock beneath the roosts where your hens sleep, digest, and poop all night. Add a couple inches of litter inside to prevent sticking, and you can empty the container every day or every few days, depending on how many chickens you have and how tidy you are.

+ Consider adding a floor in the coop that can be lowered into a diagonal position so that the litter can tip directly out and into a cart or onto the ground. This feature would be easiest to add to a coop that you are building from scratch, but you could retrofit one onto an existing coop if you are handy.

USING SAND *for* POOP MANAGEMENT

Some of our customers swear that a regularly sifted bed of sand is the poop management system of the future. One such devotee recently explained to us how she made the switch by clearing out the straw in her run and putting down a 5-inch layer of sand obtained from filled sand bags over an existing screen of ½-inch hardware cloth. Like a very large litter box, she sifts the poop from the sand a few times a week with a scooper designed to clean reptile cages. She insisted that it was far easier than other methods she'd tried and was delighted that it also eliminated the rodents that she said liked to hide in the straw. Other than scooping, maintenance consists of simply adding a few bags of sand from time to time to refresh it. After a year of managing her run litter this way, she reported that there was no odor problem and that the local health inspector gave her setup rave reviews.

Though her experience has been overwhelmingly positive, we've heard from a few other folks who have tried sand and experienced fly and odor problems, presumably caused by poop that's been washed down beyond the reach of the scooper. For this reason, we advise anyone interested in this method to limit its use to dry, covered runs and remain diligent about keeping it cleaned.

+ Slide-out floors make cleaning easier if you have a small coop. In larger coops, the weight of the litter would be substantial. You'll need plenty of clearance behind the coop to be able to slide out the long tray.

DISCOURAGING RODENTS

Rodents are a big concern for chicken keepers living in urban and suburban environments. There is a definite negative and widespread perception that this hobby invariably leads to rodent infestations. Although it is true that careless practices can lead to rodent population spikes, by taking some simple precautions, you will dissuade them from visiting your run and coop. As always, prevention is the best policy.

Tips for discouraging rodents

+ Keep all feed in rodent-proof containers, which can be as simple as metal trash cans with tight-fitting lids. Keep the feed containers at a distance from your chickens, or keep them inside the house or garage.

+ Provide only as much feed as the chickens can eat in a day so that it does not accumulate and attract unwanted diners. Or, better yet, use a feeder that prevents rat access, such as a treadle feeder, which provides feed via a platform-operated treadle that is activated by the weight of a hen.

Rodents are drawn to locations that provide access to food and water.

+ Water is another powerful rodent attractant, particularly in hot or dry weather. Use nipple-fount waterers instead of conventional double-walled founts or open bowls. Because hens can reach higher than rats and mice, suspend the waterer at the height of your chickens' beaks, and position it so that the top of the waterer is inaccessible to rodents climbing adjacent walls.

+ We urge you not to attempt to poison rodents near your hens. It's not very effective outdoors and can lead to accidental poisonings of your chickens, wildlife, and neighborhood pets. If you must, use a proper bait box and select the newer, more powerful baits. Paradoxically, these baits are considered safer because they require smaller doses to kill, whereas older types must first accumulate in the tissues of the target, potentially slowly poisoning predators such as owls, coyotes, pets, and other urban hunters that would have naturally put a dent in rodent populations.

+ Alternatively, instead of poisoning rodents, try other methods. A variety of drowning traps and other clever trapping instructions are available online if rodents are drawn to your yard. Many rodent repellents, from predator urine to ultrasonic noise generators, are marketed as safe alternatives to baits and traps, though we have yet to find one that works.

+ It's been suggested that planting mint or other strongly scented herbs near the coop will send rodents and other vermin packing. Alas, we've also found this ineffective, perhaps because the mint needs more thyme (pun intended) to grow!

PROTECTING YOUR HENS FROM THINGS THAT GO CHOMP IN THE NIGHT

Our approach has always focused on keeping our hens as safe as possible by day and, more critically, at night. Like most cities in America, Portland is home to a murderers' row of predators that emerge at dusk to prey on backyard hens and other small pets. The most notorious nocturnal consumer of chickens in our area is the ubiquitous raccoon. Fearless, dexterous, and strong, raccoons are masters of the silent night raid. Their most devious trick is reaching through small openings in wire fencing to snatch sleeping hens from their perches and pull them through the opening. If a hole is a bit larger, your hens may be visited by weasels or possums instead, with similarly grim results. Even if these guys cannot get in your

RATTING DOGS, WORKING *for a* LIVING

One summer, we brought home about forty young hens from the store to alleviate overcrowding caused by our taking in a large, wayward shipment of chicks that wound up in Portland rather than their intended destination in North Carolina. They immediately filled one shed, and as they grew, it became apparent that we would need to construct another coop for them as soon as possible. With a 500 percent increase in our already-large chicken population came a proportional increase in available chicken food—a fact that did not go unnoticed by the local rats. Their population spiked in just a few weeks—a fact that did not go unnoticed by our neighbor.

In response, we immediately ordered two (mostly) rat-proof treadle feeders and accelerated the plan to move the hens back to the store. We also reached out to Jreed and His Mongrol Hoard (sic) of Rascally Rat Wranglers—a "terrierman" with a pack of trained ratting terriers—and made arrangements for a hunt at our home on their next visit to our area. By the time they arrived, the excess hens were gone, the treadle feeders had worked their magic, and (thankfully) there weren't many rodents left to hunt. The dogs were fascinating to watch, nonetheless, as they sniffed, snorted, and energetically destroyed recently occupied tunnels in search of their quarry.

Jreed, who typically hires out his dogs only for large, working farms, sat down with us afterward and shared a few observations that will be of interest to the home chicken keeper. First and foremost, he wanted everyone to know that ratting terriers have been bred for this job for centuries and may develop behavior issues if kept primarily as pets—don't run out and adopt one, thinking that they'll solve your rodent problems. Conversely, not just any ratting terrier will make a good hunter—the effectiveness of a team like Jreed's is the result of years of hard work and training.

We were not surprised by his suggestion that chicken keepers ought to focus on prevention, but we did raise our eyebrows when he mentioned that he had seen the worst infestations in coops and runs that were fortified by concrete blocks and lots of buried wire fencing. It seems that once defenses like these have been breached (and they inevitably are), the blocks and wire become a cozy fortress for the rats, protecting them from their natural predators and preventing deep cleanings that would disrupt them. He recommended skipping the fortifications altogether and, in addition to a regular maintenance approach, once or twice a year deeply digging and flipping the soil in and around the chicken run and coop to disrupt and expose any tunneling rodents.

Specially trained ratting terriers are hard on rats, gentle on chickens.

coop, your hens may be attacked by digging coyotes, which brazenly cruise the late-night streets in many parts of the country looking for a backyard with tasty takeout. It's our duty keep our hens off the menu, but we urge you to consider the futility of poisons and traps, which only add to the suffering. As with health, prevention is always the best policy.

Protection is the most important feature of any chicken enclosure: it should be strong enough to keep hens inside and keep most predators out. Exposed hens are subject to hazards at any time. Sleeping hens make for easy targets, but the dangers don't end at dawn. By day, even active and robust hens may be lost to off-leash dogs, birds of prey, escape, and even theft. The safest enclosures possible are made from traditional wire fencing. A single layer of chicken wire can be chewed or ripped apart, but other livestock fencing material, such as hardware cloth, can be used to make very secure runs—though the heavier material can be expensive and difficult to work with.

Tips for protecting your hens from predators

✦ Site your coop in the open. We've noticed that chicken coops located in open areas seem, paradoxically, to deter predators better than those tucked into or adjacent to trees or woodlands. We suspect that predators feel less secure approaching an exposed coop, preferring to remain concealed as long as possible before

making their move. Even rats and mice are reluctant to cross open expanses to reach a coop, preferring to tunnel under if possible.

Pens surrounded by lightweight, easily reconfigured electric net fencing provide heavy-duty predator protection from the sides but not from the air.

+ Cover enclosures on all sides and on top to discourage climbing and airborne attackers. You can use heavy-duty aviary nets to provide cover. These nets, similar to those used by zoos to keep birds in open-air enclosures, are easy to secure with wire to a frame. Heavy fishing nets are a viable, less-expensive option, and used nets in good repair work just fine. That said, because a determined raccoon or other varmint could chew through netting if given enough time, we suggest you take extra precautions, such as doubling the layers of net, pinning down the edges around the base of the pen, or adorning the netting with noisy bells. Be careful not to position anything near the edge of the net that may encourage hens to roost for the night; they could be snatched.

+ Extend the fencing 1 or 2 feet beyond the sides and onto the ground outside the run or pen area. Then peg it down to thwart diggers. This is usually enough to convince predators to seek easier meals. Some folks add fencing under their runs or dig their fence edges deeply into the ground. Although both methods can be effective at keeping out predators, they can be difficult to install, and fencing placed over the ground in the run can prevent the hens from foraging. Our "fence skirt" approach has proven just as effective and is much easier, especially for mobile runs.

✦ Use an electric fence. Pasture farmers swear by reconfigurable electric net fences. This highly effective deterrent is more startling than shocking to wildlife, but it's perfectly safe for hens, who learn quickly to stay away from it.

POULTRY IN MOTION

After we moved, the chickens at our new home, now eight in number, had a spacious, prebuilt coop, a large enclosure, and were closer to the house. The setup was smaller than their former domain, and we no longer had a fenced yard for them to roam, but the new digs were nothing to snort at. For a time they enjoyed a spring and summer of renewed attention from the humans who frequently brought them weeds from the big vegetable garden and many other goodies from the kitchen. Their vitality was back, the eggs were again amazing, and the backyard nights were filled with barbecues and glasses of pinot noir once more. Best of all, our daughters (our youngest had since been born) had taken an interest in the hens as pets, and we were thrilled to watch them enjoying feeding and occasionally holding and stroking their favorites.

The good times came to an abrupt end as soggy winter weather returned once again. We unconsciously went back to our old habits, and interactions with our hens were again reduced to begrudging water fount cleanings, doling out morning feed, and half-hearted litter maintenance chores. There's nothing wrong with lying low for part of the year like this, and it was actually sort of instructive regarding the minimal amount of effort required to keep chickens, but as the results came in, they didn't look good.

The chickens were squawking and pacing discontentedly, feed costs were up because we weren't supplementing with scraps or forage, the coop was getting poopy much faster than normal, we had developed a minor rat problem, and we tossed bale after bale of straw into the run only to see it disappear in a muddy quagmire. Our hens' health also seemed to decline, evidenced by occasional bouts of watery, foul-smelling stools and the complete absence of the vitality we had formerly admired. And they laid no eggs. When the returning spring sun did little to rekindle our interest, we knew something had to change.

We ruminated on what we truly enjoyed about keeping chickens and what they needed from us to be healthy and to produce abundant, tasty eggs. We also forced ourselves to confront the chores we were avoiding, which opened our eyes to the ways predators and pests were taking advantage of our lazy ways. But mostly we thought about how to resolve chicken keeping with our busy lives—how might we simplify and have fun again?

Above right: A modern mobile day pen by Eglu provides sturdy wire construction.

Opposite, clockwise from top: A mobile chicken tractor surrounded by hardware cloth can help protect your hens and keep them moving, but the weight is a consideration.

Carrying handles and two-part construction make this setup a little easier to move.

Mobile coops are nothing new.

Could we recapture the golden years simply by letting the flock roam our yard freely once again? The thought of a poopy patio and a few lost plants didn't seem so bad, but the presence of a local pack of coyotes raised important concerns about our hens' safety. And we had become accustomed to collecting our eggs fresh from the nest box, rather than hunting for them in the shrubs.

There was no going back to free ranging, but we could try dusting off another idea from our past. A few years earlier, in our desperation to keep our early flock from covering the patio with droppings, we had slapped together a simple A-frame chicken tractor—a portable coop—from plywood, hardware cloth, and treated 2-by-4 boards.

There is nothing new or revolutionary about raising chickens in outdoor, semiconfined, mobile coops. It's a proven farming method that's been around in one form or another for at least a hundred years. If you've eaten pasture-raised eggs or grass-fed chicken meat, you've probably benefited from this simple approach. It maximizes animal welfare and produces wholesome food with a superior nutritional profile. Not to be confused with the vast, climate-controlled buildings of so-called "cage-free" or even "free-range" systems (a modest welfare improvement over battery cages that offers nothing to the nutritional quality of eggs or meat), farmers who pasture-raise their birds move them around outside in sheltered pens, protected and fed, yet also free to scratch at the soil, eat bugs, and graze plants.

The idea was sound, but in practice, it was hard to use. As we tried to load the hens into the mobile coop, the more unruly girls ran from us, and although we could usually lure them in with some corn, it seemed like too much trouble. So the rebels were allowed to remain at large, while two or three of the tamer hens always got parked together in the tractor. Even when we succeeded in loading in all eight hens, the tiny coop's cramped 8-by-4-foot dimensions would lead to terrible squabbling, and the hens would soon plaster the grass beneath it with droppings. We knew that building a larger mobile contraption might have done the trick, but the small one already weighed well over 100 pounds and was impossible for one

person to lift (though it could be dragged). A bigger one would require an actual tractor to relocate! Eventually, the beastly thing was relegated to duty as a secondary coop for injured hens, new arrivals, and mothers brooding chicks—and was never moved again.

Miniature versions of chicken tractors can also work well in urban settings. They're cheap and easy to build and capable of housing an urban flock of three hens in a small footprint, because their mobility makes their effective size practically unlimited. They are also fantastic for encouraging hens to focus that same destructive energy they unleashed on Hannah's blueberry bushes to do useful work in the garden. Wheel a chicken tractor over a pile of leaves, a compost pile, or even a weedy garden bed, and you'll quickly see why they call them tractors!

Looking back on our own experiences and those of others, we think the main problem is convenience. For a mobile coop to be practical, it must be easy to get hens in and out of it, large enough yet lightweight, and easy to maintain (cleaning our A-frame was not easy).

THE DAY PEN

Day pens—temporary runs—combine the advantages of both confinement and free-range foraging. A day pen is adaptable to any environment; you can locate the enclosure where you want to allow your flock to graze by day, and then return your hens to the more secure coop at night. You can purchase a day pen enclosure or create your own in a variety of shapes and sizes to fit your hens, your garden, and your needs, often using readily available materials. If you garden in neat, rectangular, raised beds, for example, you could create enclosures to fit over the beds; add some hens and watch as they help you prep the soil for planting.

You can also repurpose an existing structure such as a cold frame, dog run, tent, portable carport, or greenhouse to provide a frame for a day pen, or build a frame from scratch using a variety of materials, such as willow branches, PVC pipes, or bamboo. Several large patches of bamboo on our property are threatening to take over the yard, and we feel no regrets about regularly clearcutting them; they provide lots of tall poles that are ideal to lash together and use as frames for temporary runs covered in heavy aviary netting.

We have experimented with large pens in our unfenced yard to enable our hens to range in the grassy area between our fruit trees without worrying about them fleeing, and in our vegetable garden, we've used smaller pens that are easily moved within the raised beds. In the spring, after the hens have thoroughly weeded and fertilized, we remove the bird netting covering and use the framework to support a massive tomato planting.

Whichever type of framework material and shape you choose, you'll need to connect it all together. If you, like Hannah, learned all about knots and ropes while working on small boats in Penobscot Bay, by all means lash your pen frame together with rope via traditional nautical knots. When tied correctly, knotted connections stay tighter for longer, especially in windy conditions. For the rest of us, twisting heavy wire around the points where the poles cross is just fine.

GIMMIE A P! GIMMIE A V! GIMMIE A C!

Hannah was the first person in our family to discover the joys of working with PVC. We had been troubled by crows in our vegetable garden that were digging up seeds and eating our crops. Hannah decided to make a scarecrow and set to work in typical Hannah fashion. Not content making a regular scarecrow, she constructed a PVC skeleton with articulated arms and legs, fleshed out with thrift store clothing stuffed with straw, and mounted it all on a T-post. For a finishing touch, she stitched a rosy-cheeked, grinning face on a piece of burlap and stretched it over a foam ball. The girls loved it so much that they named it Hannah and insisted that she build another one, naturally named Robert. Not only does this life-sized garden inhabitant do a great job scaring the crows, we also suspect it is just the right size and shape to scare our local coyotes, and we've placed one near our coop to offer additional protection.

PVC is an extremely useful, affordable, and easy-to-work-with material for making many other helpful things, such as coops and day pens, cold frames, greenhouses, and a variety of accessories. We initially had some concerns about its sustainability and safety, but after researching and comparing it to other options, we were satisfied that most types of PVC are non-toxic and environmentally friendly when compared to other manmade materials. Bamboo is a natural alternative, and although it's more expensive (unless you grow it) and harder to work with, you can use it as a substitute for PVC in many designs. Other alternatives include flexible wood branches, copper pipe, and galvanized pipe.

Racks full of long, white PVC pipe are a common sight at building supply centers and hardware stores. PVC pipes and fittings are used primarily for drinking water supply lines, but they're also used in irrigation, sanitary, hydroponic, and gray water systems. The PVC NSF-61 grade is intended for carrying drinking water and will not leach toxins or pose other dangers under normal conditions; this type should be safe to use for delivering water throughout the garden and to your chickens.

Furniture-grade PVC, which is also non-toxic (and more expensive), is the best kind to use for building coop and pen structures. Several manufacturers produce

Opposite, top: Lash bamboo together using rope or wire to create a frame for a temporary run.

Opposite, bottom: Hannah's life-sized PVC scarecrow seems to deter chicken predators in our yard as well as garden raiders.

The LAWN MOA

Tractor coops made of PVC pipes and tarps are commonly used on small farms, mostly for seasonal pasturing of meat birds. The basic structures are indeed lightweight, easy to assemble, adequately protective—and really, really ugly in your yard. We upgraded the design of the mobile coop that we now use, improving its aesthetics while incorporating the features needed for a backyard coop (plus a few extra). The result we've dubbed "lawn moa," which combines its function with the Hawaiian word for chicken. Our lawn moa can also work as a cold frame to warm both chickens and soil in the spring. We cover it with plastic greenhouse panels, which act as insulation as well as allowing in light, but any heavyweight, clear plastic sheeting will get the job done. Keep in mind that plastic covers are no substitute for the security of an inner layer of fencing or heavy netting.

Our lawn moa protects our small flock. The design continues to evolve.

PVC pipes and fittings in a range of colors and sizes for building furniture and other DIY projects. It's also the best choice to resist the sun's UV rays, which eventually will, to varying degrees, make all other grades of PVC brittle.

The modular nature of PVC unleashes creativity, and you can design portable (or stationary) housing in any shape or size, from flexible structures using small pipe of ½ to 1 inch, to much sturdier structures with 2-inch pipe. After selecting a framing material, you'll need to select a 3D shape for your pen to fit your needs. Hundreds of plans are available online for cold frames, A-frames, domes, and teepees. Just keep in mind that it needs to be lightweight and easy to move around and/or deconstruct.

The easiest shape to construct is a teepee-shape wrapped in a heavy netting or fencing that extends beyond the base to discourage digging predators. Connect the bottoms of the poles to a frame on the ground to offer extra stability. Another easy form is a hoop house, like the bent-pole types used to cover gardens. We recommend adding one or more purlins (horizontal ridge poles) in addition to the frame at the bottom for rigidity and sheer strength. If you use star connectors, you can create geodesic domes, another efficient type of construction. In areas of heavy snow load or high winds, we recommend A-frame designs with steep sides made with thicker PVC. Lighter structures are potentially more vulnerable to weather extremes than traditional construction, and we recommend anchoring the structure in high wind and otherwise following the seasonal customs of protecting outdoor items your area.

At some point, you will need to bond your pipes and fittings together permanently to strengthen and complete your structure. We urge you to learn from our mistake and skip the traditional purple primer and colorful glues: these colors indicate to plumbing inspectors that plumbing has been properly installed, but for our use, they do little more than highlight every stain, streak, and drip. Look for specially made clear PVC primer and glue all-in-one. Not only will your PVC projects look much tidier, but this type of glue contains lower levels of volatile organic compounds (VOCs), making it safer and much less stinky than some other brands.

We also recommend using outdoor-rated spray paint to finish and protect your PVC designs. This will not only protect the PVC from damaging UV rays, but it gives you a fun opportunity to personalize your creation. Spray paint choices abound—from oil-rubbed bronze, to aged steel, copper, hammered metal, and more conventional colors in several sheens.

The NEED to FEED

PROVIDING NUTRITION

For hundreds of years, European chickens were a minor form of livestock that survived by scavenging food in towns and small farms. A few distinct breeds had been developed by the early 1800s, such as the squat and meaty Cornish. By the late 1800s, European chicken breeds were being mated with larger and more prolific egg-laying Asian counterparts. With their larger stature and greater laying abilities, these new breeds could produce a previously unimagined abundance of meat and eggs.

To achieve the full potential of the genetic improvements, new methods of feeding were developed that could provide the concentrated levels of nutrition the new breeds demanded. By the mid-twentieth century, the feeding of poultry had, quite literally, become a science. The new science looked to the rapidly industrializing factories for inspiration, and brutally efficient systems of high-density confinement factory farming were developed. When these methods reached their limits, antibiotics were fed to chickens at nontherapeutic levels—even in the absence of disease—to increase the birds' size.

The poultry industry and regulatory agencies such as the FDA in the United States have at long last recognized that this approach can lead to serious dangers such as antibiotic resistance, and researchers are now identifying a range of natural alternatives that can be used to improve poultry health while maintaining yields.

As pet chicken keeping became a popular hobby, first in the United Kingdom and later in the United States, feeding methods were initially borrowed from the farm. Paralleling the recent introduction of highly specialized cat and dog food formulas, however, chicken foods for small flocks have as of late become more diverse, offering a range of options to suit the owner's budget and preferences. Can senior and weight-control formulas be far behind?

A flock of Silver Laced Wyandotte share treats.

FEEDING FUNDAMENTALS

From the moment we bring home a box of tiny, cheeping chicks, we are responsible for their care and feeding. From chicks to the awkward teenage years, your peeps and pullets will need a specialized blend that will meet their needs for growth. The main differences between these starter, or grower, formulas and their adult counterparts are principally the amounts of protein and calcium they provide.

Just as the nutritional needs of a chick differ from those of an adult hen, breeds that are heavy layers have their own special nutritional requirements that differ from those of specialized meat breeds. Beyond age and genetic differences, numerous individual factors such as sex, climate, local weather, and overall health conditions individually and collectively influence the nutritional needs of your chickens. Your value system as a chicken keeper also plays a role in determining how best to feed your particular birds. Because it would be impossible to account for every such variable, we have narrowed our feed focus to hens of laying and ornamental breeds and assumed a temperate climate and average suburban milieu with a small lawn and garden.

So-called dual-purpose breeds (chickens that produce both meat and eggs) are the workhorses of the backyard, selected over many generations to be sturdy foragers, unafraid to get their beaks dirty and earn their keepers a yield from a few handfuls of feed and a patch of weedy ground. Most of these breeds can still be expected to lay 125 to 250 tasty eggs a year in their prime, but their laying capabilities gradually decline in production over a lifetime that may stretch to nearly ten years. The roosters of such breeds are big and meaty, if that's your thing.

Opposite, top left: Hannah's artwork adorns our latest organic feed bags.

Opposite, bottom left: A label shows the typical breakdown of ingredients in chick feed, which in this case contains 20 percent protein—more than the 16 percent typical for an adult layer feed. On most feed bags, the amount of nutrients and type of feed are usually found on the tag stitched along the top or bottom seam.

Below: Battery cages make feeding and housing laying hens cruelly efficient.

This feed is designed to be fed to layer chicks from day old up to 16 weeks...

GUARANTEED ANALYSIS

Crude Protein	Min.	20.0 %
Lysine	Min.	1.0 %
Methionine	Min.	0.32 %
Crude Fat	Min.	2.5 %
Crude Fiber	Max.	6.0 %
Calcium	Min. 0.75 % Max.	1.25 %
Phosphorus	Min.	0.5 %
Salt	Min. 0.25 % Max.	0.6 %
Sodium	Min. 0.16 % Max.	0.24 %
Manganese	Min.	110 PPM
Selenium	Min.	0.3 PPM
Vitamin A	Min.	3,300 IU/LB
Vitamin E	Min.	1,000 IU/LB
Vitamin D3	Min.	20.0 IU/LB
Lactobacillus acidophilus	Min. 10.2 m1 CFU/LB	
Lactobacillus casei	Min. 10.2 m1 CFU/LB	
Bifidobacteria termofilum	Min. 10.2 m1 CFU/LB	
Enterococcus faecium	Min. 10.2 m1 CFU/LB	

INGREDIENTS: Organic Ground Wheat, Organic Soybean Meal, Organic Middlings, Organic Ground Corn, Calcium Carbonate, Soybean Oil, Dicalcium Phosphate, Monocalcium Phos Salt, Citric Acid (a preservative), Dried Penicillium Fermentation Extract, Sodium Bicarbonate, Choline Ch Methionine, Selenium Yeast, Vitamin E Supplement, V Supplement, Vitamin D3 Supplement, Vitamin B12 Su Riboflavin Supplement, Niacin Supplement, d-Calcium Pantothenate, Pyridoxine Hydrochloride, Folic Acid, M Sodium Bisulfite Complex (source of Vitamin K activit Mononitrate, Biotin, Ferrous Sulfate, Manganous Oxid Limestone, Zinc Oxide, Sodium Selenite, Basic Copp Ethylenediamine Dihydroiodide, Yeast Culture, Orga Hulls, Organic Cane Molasses, Organic Rice Mill By Dried Bifidobacterium thermophilum Fermentation Pr Lactobacillus acidophilus Fermentation Product, Drie Lactobacillus casei Fermentation Product, Dried En faecium Fermentation Product.

FEEDING DIRECTIONS: Feed this complete feed continuously as sole ration to starting pullets, breeders and broilers. Layers: Feed from day old up to 16 weeks of age for layers. After 16 weeks, gradually switch older birds over to Natures Layer Pellets or Crumbles. Broilers: Feed free choice from hatching until birds are marketed. Keep feeders and waterers clean and disinfected to h...

Far right:
Orpingtons may look frivolously fluffy, but they are a solid heritage breed.

By contrast, more recently developed "super layers" are engineered for a sixteen-month life at a commercial farm. These hens are devoted solely to laying while locked in a cage and efficiently produce about 350 eggs a year. If they are so different, why should we feed them the same way we feed our backyard hens?

Open an agricultural textbook and it will tell you that a well-provisioned laying hen needs precise amounts of water and finely tuned inputs of energy in the form of carbohydrates and fats, balanced by protein sufficient for growth and egg production. In addition to these, the ration must include minimal levels of calcium, phosphorus, sodium, and other vitamins and minerals to prevent deficiencies and provide for the birds' basic needs during their short lifetimes.

We believe that this deconstructionist approach to poultry nutrition is best suited to chicken farmers, who are governed by cost and efficiency. The backyard chicken keeper's concerns lay primarily in the flock's well-being and the quality of the eggs they produce. Backyard chicken keepers can do much better than provide nutrition that meets the minimum requirements needed to keep a bird alive; we believe strongly in the importance of offering access to living foods such as weeds, grasses, bugs, and soil microbes. Nevertheless, research for commercial chicken farms has been done, and redone, and provides useful information that can help guide us home keepers as well by serving as a baseline against which we can judge the completeness of the diets we provide our hens.

ON THE HUNT

Chickens are omnivores, equipped to eat and digest a wide variety of plants and animals to meet their nutritional needs. Given a hospitable climate and lack of predators, chickens are able to adapt to a variety of locally available food resources. By constantly digging around, nibbling, and exploring, the feral Kauai chickens scratch together a diet complete enough to sustain them in top condition, supplemented only by the occasional French fry snatched from a tourist's plate. But they lay only about 15 eggs per year.

Our domestic chickens retain a surprising amount of this versatility, and one of the great joys of being a chicken keeper is watching our hens find food in the yard. Unrecognizable as the plump, lazy ladies squawking at us to refill the feeder only moments earlier, our foraging hens see something interesting and turn into stealthy raptors on the hunt. Their long, slow strides enable them to listen and look at once, sensing subtle vibrations with their feet that betray the presence of the quarry hidden below. When something promising is located, they begin methodically peeling back layers of duff and leaves with their powerful feet and sharp claws, pecking and bending, pushing soil with their beaks, and claiming bit after bit of something usually invisible to our clumsy human eyes.

The other sort of feeding style our hens use begins with a tentative tip-grazing of plants. If they find it tender and delicious, they tuck in with zeal. It's astonishing to witness a couple of hens devour a full-grown kale plant in about ten minutes, reducing it to a roughed-up patch of mangled roots and a few of the tougher stems. Three or four hens in a foraging pen can completely clear a moderately weedy 8-by-4-foot garden bed in a day or two, leaving it weeded, fluffed, and fertilized for the next crop.

BAG IT, TAG IT

As satisfying as it is to watch and assist the hens with their foraging (by moving the pens, planting greens for them, and performing other such favors), and despite our fantasies to the contrary, we must provide the bulk of their food in

On the hunt for a tasty worm in the pasture

a more complete, consistent, and concentrated source of nutrition if we want them to lay abundant eggs and otherwise live up to their breeds' potential. Today, about 70 percent of all chicken feed comes in pellet form. Pelletized feed is made by compacting ground and blended ingredients to form small, solid, elongated pellets. The purpose of pelleting is to take fine, dusty, and difficult-to-handle feed material and, by using low heat, moisture, and pressure, form it into larger bits that are evenly mixed to deliver the complete blend of the feed's formula in every bite.

The other form of feed is mash—loose, unpelletized feed. One popular brand offers absolutely gorgeous stuff, almost like the ten-grain cereal you eat at breakfast. You can clearly see coarsely cracked grains with a scattering of chartreuse dry pea bits peeking out here and there, adding some color to the wholesome effect. It looks just as a natural chicken feed ought to look, but there's one problem: most people use it incorrectly when feeding their hens, and this means their hens are missing out on an important fraction of the nutrition it contains. The loose structure of the feed enables the hens to pick out the bits they like the most—mainly the grains—while leaving behind a substantial portion of the dusty parts. But the dusty stuff contains the majority of the vitamins and minerals, and a portion of the protein and fats that make the feed complete. We regularly hear reports from customers who are loyal to mashes that their hens' eggs are thin-shelled or they're having other difficulties; after we instruct them on proper feeding techniques or switch them to pellets, the problems disappear. The feeding issue is shared by all feeds with a lot of variation in particle size—including most homemade feeds—but it's not a problem with fine mashes that are sometimes fed to young chicks, which are homogenous enough that the birds will ingest it all.

To avoid inconsistencies when feeding loose feed with a variety of particle sizes, we recommend mixing up an old-fashioned wet mash—not very wet, mind you, but just moist enough to form a stiff paste that binds the ingredients together. The amount of moisture will vary by feed type and composition, but a five parts feed to one part moisture (by volume, not weight) will probably do. This is also a great first step toward fermenting your feed if you desire. Farmers in the past may have added a little squirt of milk from Bessie to get this done, and you can use milk, too—but keep in mind that Bess would have given raw milk that offered far better results than the pasteurized type. So it's probably safer to go with water, with a little apple cider vinegar or humic acid mixed in to keep the pH on the low side. When in doubt, read the label on the feed, follow directions, and judge for yourself.

Using a wet mash, however, has two major downsides: preparation and mess. It will take a little extra time to mix the warm water with the feed, a little more time to let it soak in, and up to two days for fermentation to occur if that's your aim. Then you'll need extra time to clean up after the hens' previous day's meal, because any leftovers must be removed from the feeder and the feeder rinsed (or sanitized, if you are really tidy). Keep in mind that only a trough feeder is suitable for wet-mash feeding, because the mushy stuff won't flow in gravity-fed feeders.

THE ELEMENTS OF NUTRITION

Nutritional needs and restrictions gradually change as your chicks age, making the task of selecting the appropriate feed, treats, and supplements daunting. This handy chart summarizes the type and amount of feed your chickens require at each stage of growth and the corresponding recommended treats, minerals, and supplements that can safely be given daily.

From chick to adult, laying hens can be expected to thrive on nothing more than an age-appropriate feed formula and an occasional nibble of green forage. Perhaps because we find this monotonous, we pamper our flocks with a variety of supplemental treats, especially cracked grains. The variety these provide can reduce boredom and may improve digestion, but, because they displace consumption of complete feeds, moderation is essential. Mineral additions such as grit and oyster shell are less prone to abuse and may simply be scattered in small amounts over feed or on the ground.

Opposite: In some rural areas throughout the world, chickens are fed whatever scraps are on hand.

DAILY FEED AND SUPPLEMENTS BY AGE

Life stage (age)	Base feed *required*	Treats and greens *optional*	Minerals *optional*	Botanicals and microbes *optional*
Hatch to 8 weeks	Chick grower 20% protein (unlimited)	Mealworms (no earthworms, may contain parasites dangerous to chicks), hardboiled egg bits, polenta/chick corn, small bits of meat	Fine grit, bentonite clay, biochar	Aloe, artemisia, probiotics, bokashi bran or EM/BM, kelp
8–12 weeks (transition to outside)	Chick grower or developer 18–20% protein (unlimited)	Mealworms, scratch grains, leafy greens, tender grass	Fine (or larger) grit, bentonite clay, biochar	Aloe, artemisia, probiotics, bokashi bran or EM/BM, kelp
12 weeks to adult layers	Layer mix 16–17% protein, 5–6 oz. daily	Mealworms, scratch grains, leafy greens, tender grass, limited kitchen scraps including veggies, meat, cheese, bread	Regular grit, bentonite clay, biochar, calcium	Aloe, artemisia, probiotics, bokashi bran or EM/BM, kelp

CARBOHYDRATES

Carbohydrates are the major source of energy for poultry and the largest component of their feed by weight and volume. Carbohydrates consist of starches, sugars, and fiber. Although starches and sugars are easily digested by enzymes in a chicken's digestive tract and are essential for fueling a hen, the less digestible fibrous parts in proper proportion are also important for maintaining digestive health. Corn is the major source of carbohydrates found in most milled feeds because it's an inexpensive and high-energy grain, but some mixes contain other cereal grains, including wheat, barley, rice, millet, quinoa, and amaranth as well. Grains also contain protein, but usually in amounts insufficient to provide for a hen's total protein needs. Therefore, they must be combined with more concentrated protein sources to yield a complete ration. Feeding your hens excess carbohydrates (usually in the form of treats) will lead to serious health problems, including obesity and fatty liver disease. Too little in the diet can lead to low body weight.

BENEFICIAL MICROBES *and* EM

Effective Microorganisms (EM) is a proprietary blend of beneficial microorganisms first cultured in Japan by researcher Dr. Teruo Higa in 1982. Teraganix is licensed to use Higa's recipe and the EM name in the United States. According to the Teraganix website, the microbe cohorts in EM "work together with local and native beneficial microbes, creating a synergy among microorganisms and larger forms of life including insects and worms, pets and livestock, and people."

To avoid confusion (and lawsuits), we have recently begun referring to EM and similar products as beneficial microbes, a less specific but less legally protected name, and the slightly tongue-in-cheek acronym, BM.

The specific blend of microbes present in EM and other commercial BM formulations are shrouded in secrecy, but they are known to contain specific strains of lactic acid bacteria, yeasts, and phototrophic bacteria. Several recipes exist for homebrewed BM starters, but the processes are tricky enough that we suspect the results would be inconsistent for most of us.

DIGESTIVE ENZYMES DEMYSTIFIED

Digestive enzymes are molecules produced within the digestive systems of humans, chickens, and other animals. They efficiently break apart chemical bonds in food to make it digestible. Once the enzymes transform food into smaller molecular blocks, the molecules can be absorbed and used throughout the body. In chickens (as well as humans), for example, the amylase enzyme breaks the long molecular chains of starches into sugars, and pepsin breaks down protein molecules.

In addition to the enzymes created inside the body, many are helpfully present within food itself. Chickens (as well as humans) cannot digest some foods because their bodies lack the enzyme required to do so; for example, the enzyme cellulase, which is not produced by chickens (or humans), is required to break down plant fibers containing cellulose. Enzymes are sometimes added to feed to help chickens digest grass and other cellulose-rich green plants. Adding selected enzymes to poultry feed improves its digestibility. The casual chicken keeper need not concern herself with the details—it's enough to know that enzymes unlock the full nutritional potential of feeds and it's crucial that they be included in the blend and consumed with the other ingredients.

Corn This grain (properly called maize) deserves its reputation as an overabundant industrial crop that's often refined into dangerously addictive human food additives such as high fructose corn syrup. Nevertheless, it's not without merit as a source of carbohydrates for chickens. Unlike ruminants (cud chewers) such as cattle, which were never meant to digest large amounts of low-fiber seeds such as maize, chickens do just fine on the stuff. In fact, maize is one of the most energy-dense foods for poultry, with a digestible energy (DE) value of more than twice that of oats. Maize weighs in at a hefty 1.54 megacalories of DE per pound: that's equal to 1540 calories of the kind people are concerned with (which are actually kilocalories). To put that in perspective, 1 pound of dried maize alone would nearly meet the basic daily energy requirements of our five-year-old daughter!

Although we don't use maize in our retail organic hen food blends at this time, we do appreciate it for being a relatively inexpensive source of energy. Our milling partners probably wish we would use it for another reason: its low fiber-to-energy ratio means that it's concentrated, leaving plenty of volume in feed formulas for other, bulkier ingredients. Its greatest virtue for the home chicken keeper is probably as a crop to grow at home to save a little money or to meet food security and sustainability aims. In addition to planting one of the many easy-to-grow traditional feed maize varieties, you can also supplement your hens' diet with

Chickens will gladly eat entire ears of corn, fresh off the stalk.

GROW *your own* HEIRLOOM CORN

Our daughters, now ages five and seven, begin begging us for corn with dinner in early July each summer. Until our own crop of ultra-tender and sweet corn begins to ripen in August, they endure the store-bought kind, which they deem acceptable if sufficiently slathered in butter.

When they're old enough to appreciate this sort of fascinating information, we'll share the surprising fact that sweet corn eaten fresh is, in fact, a rare use of corn in the United States. Instead, most owf our corn crop is distilled into biofuels such as ethanol (about 40 percent), and almost all the rest is used as animal feed (about 36 percent, or a bit more if distillers' grain left over from biofuel production is included). Most of the rest is exported, and only a tiny fraction of our nation's corn crop is used directly for human food—and primarily consumed as high-fructose corn syrup.

If our daughters are able to contain their excitement long enough to hear more, we'll teach them about different kinds of corn as well. We can thank a naturally occurring mutation in genes regulating the conversion of sugar to starch inside the maturing kernels for its sweet taste. Corn without this mutation is known generally as field corn. This corn is used to make ethanol, corn meal, and animal feed. Home gardeners typically grow sweet corn, but field corn can also be cultivated easily in most parts of the country. Easily dried and stored, field corn makes an ideal home crop for stretching your flock's food supply.

Specifically bred for today's modern markets, common agricultural corn varieties bear little resemblance to the largely forgotten, older types better suited for the home gardener interested in feeding chickens. Although hardly heirlooms, modern corn types available to the public are hybridized versions of a respectably old type of maize, called dent corn, which was developed in the mid-1800s. The original corn varieties used to create the new ones didn't completely disappear, however, and it's still possible to grow a few of them, as well as many more traditional and newer variants, at home. With a little scratching around, you should be able to dig up such colorfully named dent corn varieties as 'Bloody Butcher', 'Blue Clarage', 'Cherokee White Eagle', and a rainbow of others.

Flint corn is another distinct type with a long history and a good candidate for the home garden. Better garden-grown sweet corn that's gone starchy or dried in the husk. We like to save a few dozen leftover ears, dry them thoroughly, stash them in a rodent-proof bin, and offer them whole as a sweet treat in the depths of winter. Homegrown popcorn and other colored corns are also perfect for feeding to your hens.

Because hens love corn's sweet taste, they will always eat corn before their pellets or regular mash, which can result in decreased protein and mineral intake. So be cautious about feeding chickens too much. It's better to include cracked corn as an adjunct to your winter ration. Your hens' protein needs decrease when laying slows in the colder months, but their energy demand increases to maintain body temperature. In mild winter areas, try giving them foraged green plants rather than stored corn for supplemental food.

A different kind of corn, perfect for chicken feed

suited than dent corn for gardens in cooler climates, flint also comes in a charming assortment of colorful options and makes a quality feed or scratch. Though you may choose from enticing variety names such as 'Floriani Red', 'Saskatchewan Rainbow', or 'Wade's Giant', flint corn seems to be more difficult to find than dent corn—perhaps indicating that it's losing popularity and is worthy of our preservation attention.

Popcorn is wonderfully suited for home cultivation and is ideal as a long-storing food for both you and your flock. Numerous old and new varieties are readily available to home gardeners, including the hardy black ears of 'Dakota Black' and the pastel colors of the opalescent variety 'Glass Gem'. Try growing 'Pennsylvania Dutch Butter Flavor' to see if it really tastes prebuttered!

Sweet corn is not bred to be dried before eating and contains lots of sugars and water. Dent, flint, and popcorn varieties are much starchier. Although it's perfectly fine for feeding to chickens, sweet corn shrinks dramatically when dried, and this makes it impractical to grow expressly for feeding chickens.

Whichever type you grow, you'll be harvesting not only a valuable food for you and your flock, but you'll be doing your part to maintain genetic diversity, much as you probably already have by choosing heritage and dual-purpose chicken breeds. We've recently begun growing open-pollinated, non-hybrid field corn in our garden and plan to try a new variety each year to determine which is best suited for our location. Perhaps we will even cross two or more promising strains and try our hand at creating something unique.

Wheat Many of our customers' interests are piqued when they see wheat included in our feed mixes. They are concerned about the environmental impacts of corn and soy, but we typically formulate our proprietary hen foods without either. Instead, we use wheat as the primary energy source. Wheat offers a few advantages over corn as a source of carbohydrates in poultry feed, including a higher protein content and a wider range of amino acids, which are essential in every metabolic process inside a hen's body. These benefits are mitigated somewhat by wheat's lower energy content and indigestible starches—unless appropriate digestive enzymes are included in the ration as well. By blending wheat with naturally produced, organic enzymes such as xylanase, protease, and amylase, its energy value is unlocked and relates favorably to that of corn. Another advantage is that the natural gluten in wheat eliminates the need for artificial binders because it helps to bind feed pellets, so bags of pelletized feed are much less dusty.

Wheat is often added to feed rations in the form of wheat byproducts, such as middlings. When we first started to formulate feeds, we had an aversion to the idea of adding byproducts such as middlings because they sounded like low-quality ingredients, filler, or, worse yet, something that was unnatural for hens to eat. We've since learned that, although that may be true in a few rare instances, most grain byproduct ingredients are actually very desirable in feed. They are the part of a whole food that's removed during processing, or refinement. Wheat middlings, for instance, are simply the remains of whole-wheat kernels that have been milled into white flour. They consist of the germ and the bran, the parts that are actually richest in proteins, vitamins, lipids, and minerals.

Wheat is also a good forage product in the form of fresh green wheatgrass, but unless you are willing to thresh it and include digestive enzymes with your supplements, growing wheat for grain is not very practical for most time-crunched urban homesteaders like us. It may be an attainable goal to use homegrown wheat in a scratch blend that is threshed and cracked on a small scale, however.

Amaranth This highly nutritious grain produces large yields in small spaces, making it ideal for an organic feed and forage garden. An exotic-looking annual plant that can be grown in most climates, amaranth produces large drooping

GLUTEN SENSITIVITIES

People who eat gluten-free diets have asked us whether feeding wheat to their hens will give them a reaction when they eat the eggs. The short answer is no. A hen's digestive system breaks down the gluten into amino acids that retain zero of the potentially harmful or irritating properties of whole gluten proteins.

The long explanation is that after your hen swallows food containing gluten, a series of chemical and mechanical processes begin to reduce it to a form that can be readily absorbed by her body and used for nourishment and growth. As the food undergoes this process, proteins such as gluten are broken down into amino acids. Unlike large and complicated protein molecules, amino acids are found in almost all types of foods in one form or another, and they are tiny enough to be absorbed through the wall of the small intestine.

We've been asked this question so many times that we've researched it thoroughly. We've been unable to locate any possible mechanism whereby the gluten protein could remain whole and potent and then be incorporated into an egg. Even if somehow a gluten particle escaped digestion, it would join other undigested bits and simply end up in a hen's poop.

POTATOES, *you* SAY?

As a source of nourishment for a small flock, home-grown potatoes are easy to cultivate and highly valued for winter chicken feeding—but they must be cooked before you give them to your hens. They are a good source of energy for poultry and provide moderate protein levels, though this varies by variety and cultivation method.

Potatoes and their byproducts contain a naturally occurring antinutrient, which blocks proper digestion of proteins by inhibiting the action of the digestive enzyme protease. Because of the protease inhibitor, raw potatoes are unsuitable for feeding to hens. Cooked potatoes, however, are quite safe, and it's critically important that you cook them completely before feeding them to your flock. Mash the cooked potatoes and the feed together, and place the mix in a trough other than the one used for their regular dry food to prevent moisture transfer that would lead to spoilage. Our hens prefer this mash to regular pellets, but if they didn't, we might find it necessary to remove the regular feed so they would focus on the potato mix.

We tend to be cautious and limit cooked potatoes to 20 percent or less of our hens' diet, though we've seen evidence supporting the safety of higher percentages. Potatoes contain less protein than milled hen food pellets, so swapping them for 20 percent or less of a day's meal will mean a slight decrease in protein consumption. To balance it out, we toss the girls a few worms or another high-protein supplement. Otherwise, a day or two of missing the recommended nutritional marks will not noticeably affect hens' health or productivity.

We dedicate a substantial portion of our vegetable garden to several types of potatoes, and we are constantly amazed at yields: for each pound of potatoes we plant in early spring, we harvest ten or more pounds by the end of the season. To get those outstanding returns, we use hilling techniques, add low-nitrogen organic amendments, always plant certified and disease-free seed potatoes, and resist the temptation to use the sprouted spuds we find at the back of our cupboard. After a season of growing in most rich garden soils, replanted potatoes are saddled with a load of pathogens that rob them of some vigor and reduce yields. Seed potatoes are grown in thin, dry soils.

When stored hanging in burlap bags in a root cellar or cool, dark garage, spuds will last long into the winter. Potatoes with green coloration should be avoided because they contain solanine, which is toxic, though it's less bothersome to chickens than humans.

clusters of tightly packed seeds that ripen to attractive shades of yellow, purple, and red. The seeds are rich in protein (18–20 percent), contain beneficial amino acids, and offer higher energy levels than most cereals. Amaranth also includes unsaturated fatty acids that are important for poultry growth, egg production, liver health, and immune function. Amaranth's tiny seeds and tough husks make it difficult to thresh for human consumption, but chickens with access to grit can digest seed and hulls. Its antinutritional factors, similar to those in some other raw grains, require that amaranth seed be heated or fermented to help hens digest it. Amaranth leaves are also rich in protein (up to 25 percent of dry weight) but must be dried before feeding to improve digestibility.

Barley Residents of Canada and Northern Europe have shown that barley can be fed to chickens, though chickens cannot digest its carbohydrates as easily as those of corn. Barley provides moderate amounts of energy and lots of fiber, but it must be supplemented with enzymes to negate compounds such as phytate, which can slow absorption of other nutrients. In brewing and distilling alcoholic beverages through the malting process, barley is partially spouted and then quickly dried to convert starches within the grain into sugars. The complex sprouting and drying process improves the grain's digestibility by releasing enzymes that make barley a more viable ingredient in feed. Barley is often available for free in the form of spent brewers' grains from breweries and distillers, but don't seek it out for your flock; although this is suitable feedstock for multi-stomached, cud chewing, cellulose-digesting ruminant animals such as cattle, chickens can make little use of it. This leftover material will contain mostly fiber and fat, with no sugar.

Millet Millet is a cereal grass grown in difficult conditions, such as those found in many food-insecure parts of the world. With high levels of protein (20 percent) and as much energy as corn, millet could practically be the main ingredient in chicken feed, requiring only minor supplementation. Before you get too excited about switching to an all-millet diet, though, we should mention that it's prone to fungal attacks that can produce toxins dangerous to poultry and humans. If you use it, source human-grade millet from trustworthy suppliers that regularly test it for safety.

Oats We don't typically use oats in our poultry feeds because our milling partners have always told us that its high fiber content makes it difficult to balance with other ingredients. Although this is undoubtedly true, recent research has confirmed what pastured poultry and clean-food guru Joel Salatin has been saying for a long time: fiber is very valuable in a chicken's diet. When it's young and tender, oat grass makes a terrific grazing grass for hens. At home, we occasionally mix leftover oatmeal with our regular feed and a few kitchen tidbits for a treat.

Rice A high-energy grain, rice is becoming increasingly popular as poultry feed in areas of the world where it's abundant. Rice bran is a great byproduct of the industry, with an excellent nutritional profile rich in valuable fats, vitamins, and minerals. Because it contains fats, rice bran spoils fairly quickly in warm climates, so it should be stored in a cool, dark place and vacuum-sealed when kept for more than a few weeks. We use rice or oat bran to make bokashi for composting our food waste and as a supplement for our hens.

Sorghum This grain may be most familiar as those little round things in wild bird seed. It's a heat- and drought-resistant grain that's loaded with nearly as much poultry-digestible energy as corn. Sorghum thrives in many parts of the world where other crops fail, but high levels of tannins (bitter compounds like those in black tea) relegate it to being a minor component of animal feeds. Recently, low-tannin varieties have been developed that should enable its use in greater amounts, making it an intriguing substitute for corn.

FATS

Fats are the richest source of dietary energy, containing 225 percent more calories than carbohydrates by weight. Fat must be present in a chicken's diet to help its body absorb fat-soluble vitamins A, D, E, and K; it also slows passage of food through the digestive tract, improving nutrient absorption. And, not surprisingly, it improves the taste of their food. Soy oil and ground flax seed are common fat sources in chicken feed. Fats can go bad or become rancid, especially in warm temperatures. A controversial preservative, ethoxyquin, has been deemed safe for animal feed, but we prefer to include fats preserved with vitamin E in our mixes.

Fats can be saturated, which are solid at room temperature, and unsaturated, which remain liquid. Both can be part of a healthy poultry diet if used appropriately, though oils containing unsaturated fats are better used in a hen's body than saturated fats. Examples of saturated fat include vegetable shortening, lard, and tallow, and although we don't typically recommend adding saturated fat to regular feeds, they are useful for making a warming winter hen suet. You can use high-quality, sustainably produced animal fats, but we recommend coconut oil as a more sustainable (if somewhat pricier) alternative. Unsaturated fats are much more commonly used in feeds, including home-mixed blends, to control dust and as a concentrated form of energy to balance, or pump up, the caloric content. The most common types of unsaturated fat used in chicken feed are corn, soy, and canola oils. Although each has its benefits and drawbacks, we generally find them all to be affordable and acceptable. Alternatives such as avocado and rice bran oils are nutritionally stellar but astronomically expensive.

Because the base ingredients vary widely in their fat content, it's difficult to provide exact recommendations about the amount of fats to add when putting together a layer ration. As a rule of thumb, rations based primarily on whole grains and (cooked or extruded) legumes have enough oil present to make a healthful ration without including additional fat. However, when using less-expensive byproduct seed meals that have been previously pressed for oil, you must add fat to compensate for this. For instance, soybeans contain about 20 percent oil prior to pressing and only 1 to 2 percent afterward. In theory, if you include soybean meal in a ration, we suggest you add about 18 percent of the weight of the meal in the form of oil. In practice, this is a bit much and would result in an oily product, but it does provide a useful starting point to begin customizing for your needs and preferences.

SCRATCH GRAINS

It's a common practice to offer treats to your hens in the form of scratch, so called because chickens will scratch the ground looking for every last tasty morsel. Scratch can comprise a single ingredient or blends of items that hens particularly relish, and it is useful for training hens to come to your call. Treats with high energy content are also a warming supplement on cold nights.

Our customers are very fond of their chickens, and, like other folks who love their pets, many of them express their love in a language their hens really respond to: yummy treats. We'll admit that it's good for business, and the hens certainly don't mind, but sometimes we are compelled to inquire gently about a customer's feeding practices. It usually goes like this:

Left to right: whole-grain mash, pelleted feed, corn scratch, six-grain scratch

"Good morning, I'd like a 50-pound bag of cracked corn treats please—the organic kind."

"Sure, we can grab that for you. Must have lots of chickens, huh? Did you need some regular feed, too?"

"No, we have only two now. We lost Gracie last month. Just found her dead suddenly. It's so sad. We've still got plenty of the regular food—we're just out of the corn."

"Always sad to lose a favorite hen. . . . Sooo, how you feed them the corn?"

"Every morning, when I go out to visit the ladies, I give them a handful. I just hang out with them, drink some coffee, and we chat away! They finish the corn really fast, so I usually toss them a little bit more. They love that corn so much. They scream if I don't give it to them—it's so cute!"

"Yeah, hens are like that. They like to talk about stuff . . . loud! So, anyway, about the regular feed, do they ever talk about that? Do they eat much?"

"Nope, they say they don't like it very much. It just sits there in the feeder. I think they eat some of it. We refill the feeder every few weeks. Maybe I need to try a different brand or something. It seems they don't like the kind I get them anymore, or they're bored of the stuff."

"Right. They do get bored sometimes. So how are they doing with the egg laying? It's getting warmer, so they must be really cranking them out now, huh?"

"The eggs? Oh, they're delicious! When we get them, that is, because sometimes they hide them. . . . Oh, that reminds me, I think Susie might be eating her eggs, and she's starting to peck at Buttercup's feathers—plucks them right off! I tell her she's a naughty girl, but she won't listen."

No longer able to contain ourselves and in real danger of bursting from the pressure of withheld advice, this is the moment we simply must launch into our famous "tough love for chickens" speech—where we firmly proclaim that hens must eat a balanced diet with the correct balance of carbohydrates and protein to supply energy, maintain and repair their bodies, and produce lots of eggs. Sugary corn and starchy treats like bread are high in energy but low in protein. They may seem alright because they're wholesome grains and chickens really like to eat them, but the reality is that feeding hens too many treats can kill them, or at least keep them from reaching their potential. Inexorably, feeding too many treats leads to protein deficiency—an underlying cause of serious problems such as egg eating, feather picking, cannibalism, aggressive behavior, and declines in egg laying. The excess carbs may disrupt the balance of microbes in the gut for the worse, opening the door for pathogens and the diseases they cause.

If, by this time, the customer has not suddenly excused herself to make a phone call or respond to an emergency at work (this is always happening to us for some reason), we'll go on to explain that the trouble starts because chickens will eat to their food's energy level, meaning that they will consume all of their food intake as carbs, if that's available, and neglect to eat offerings containing protein. "It's like giving kids candy before breakfast," we say wisely, "or how they always have plenty of room for dessert, even after they say they're full of our famous liver, onion, and cheddar fondue. You've got to get firm with your hens! What they need is some tough love. Stop giving them treats for a few days and make them eat their pellets—that'll set them right." About this time, we finally notice that the customer's face is frozen in a look of concern. "Don't worry, ma'am, they'll get hungry and eat the pellets . . . sooner or later," we add, reassuringly.

Flax seeds Valuable as both a protein and an oil, flax is rich in omega-3 fatty acids. When fed to laying hens, feeds containing sufficient levels of flax oil are known to produce changes in the fat profile of the egg that may lower cholesterol in humans and provide other health benefits to the birds. Unfortunately, our preliminary research indicates that to achieve a significant bump in omega-3 levels, it may be necessary to use up to 10 percent flax in a feed blend, which would add substantial expense. A more economical choice to boost omega-3 levels, and one that we've been incorporating into some of our feed formulas, is fish meal. Unfortunately, mills are moving away from its use because of generalized concerns about using animal proteins in feed—and perhaps because of its odor.

Hemp seeds Like flax seed, hemp seed is rich in quality fats. Hemp can boast good levels of energy and is a superior source of plant protein, with an excellent amino acid balance. The seed is so nutritious that it could be a one-ingredient feed, comparing favorably with modern, blended chicken feeds.

Sunflower seeds If you have access to a small oil press, you can make your own oils from seeds. The easiest oil to make comes from sunflower seeds, and this plant may be growing in your garden already. We recommend using a small-seeded variety, the black oil sunflower, which is often included in wild bird seed mixes.

PROTEIN

Hens need protein in their diet for growth and repair of bodily tissues and feathers. It's also an important factor in egg formation. Proteins are composed of amino acids and vary in their completeness, or usability. Supplementing protein intake periodically can boost egg production and speed the replacement of feathers during molting. Eggs may be fed as supplement to your chicks' and hens' diet if they are cooked and unrecognizable (no, it's not cannibalism— eggs are the food supply for chicks before they are hatched, after all). Typical sources of proteins in feed include soybean meal (a byproduct of soy oil production) and animal byproducts from beef, swine, and (shockingly) chickens. Also common are various byproducts of seeds pressed for oil, such as canola. Sunflower meal, fish meal, hemp meal, flax meal (linseed), insects, and grubs and maggots can also be fed to hens to provide protein. Hens deficient in protein exhibit slow growth, decreased egg production and size, earlier and prolonged molting, feather picking, aggressive behaviors, and egg eating. An excess of protein shows up as enlarged eggs that cause potentially fatal binding, increased water needs, and elevated ammonia levels in litter.

Opposite:
Sunflower seeds are an easily grown source of protein.

Below:
No gumballs— just protein-rich mealworms as treats for our store chickens

Sunflower seeds An ideal ingredient for the home feed maker, sunflower seeds also make a good scratch treat. The black oil varieties are highly productive and produce seeds that are stable in storage, and their small size makes them easy for the hens to eat. Other varieties are fine, too. Keep in mind that you may need to net your sunflower plants as the seeds become ripe to protect them from being eaten by other birds, and with very tall varieties, this can be difficult. Our scarecrows protect our sunflowers pretty well, but we like to leave all of the smaller heads on the stalks for wild birds anyway.

We currently use whole sunflower seeds to provide as a scratch treat, but if you want to get serious about creating your own feed mixes, consider buying or going in with a group of other like-minded chicken keepers in your area on a hand mill that's capable of shelling them.

EGG EATING

One of the most vexing and frustrating characters a chicken keeper can grapple with is the egg-eating hen. It begins with an innocent nibble of raw egg from within a cracked shell in the nest, but once she's tasted the yummy things inside, she'll become an egg fiend. Follow a few prevention and treatment tips to help keep this problem at bay.

• Collect eggs dutifully each day to remove temptation. Do this as close to the time of laying as possible, typically late morning.

• Leave dummy (fake) eggs in the nest to serve as decoys, possibly deflecting pecks and making all eggs seem less appealing once the hard, impenetrable substitute fails to yield anything interesting.

• Maintain shell quality and density by feeding your hens proper levels of calcium, vitamins, and protein to help prevent unintentional breakage that could lead to taste tests.

• Use roll-out nest boxes with sloped bottoms to deposit newly laid eggs safely away from the hens in a protected compartment for later gathering.

• Provide a sufficient number of nest boxes for the flock size.

Once a hen gets a taste of a fresh egg, she's hooked. It's a habit that's difficult to break.

Sunflower meal is one of our preferred alternative sources of protein sources (along with peas) for several reasons.

✦ It contributes to agricultural diversity as a protein alternative to industrially farmed soy beans.

✦ Sunflowers are native to North America, which means they are well adapted to our climate and soils, reducing the needs for water, fertilizer, and pesticides.

✦ Sunflowers are virtually free of antinutrients, which are present in soy.

✦ They provide a well-balanced amino acid composition for quality protein.

The nutritionist at the mill that makes our feed tells us that one drawback to relying on sunflower meal as a primary protein is its high fiber content. Compared to soybean meal, hens must eat a bit more sunflower meal to meet their protein requirements, making the feed formula less nutrient-dense overall. This is not a problem for feeding home flocks, however, because the extra fiber may be nutritionally useful for confined chickens with low plant fiber diets.

FORAGE: GREENS AND GRUBS

In our experience, chicken keepers are almost invariably gardeners as well. Whether you are the casual sort, keeping a few layers and managing a harvest of homegrown tomatoes in a small garden, or folks like us, whose gardens are actually farms writ small, complete with hens as scaled-to-fit livestock (and pets), you can use your plot to obtain a better yield of eggs while spending less on feed. Though forage is not a replacement for regular feed, it can provide at least 10 percent of a flock's nutritional needs, depending on seasonal availability and your skill at providing it. Too much foraged food and your hens will likely consume a diet lower in overall nutrients, by effectively diluting the nutrient density of regular feed.

Foraging for both plants and insects is the natural behavior of chickens. Whether gathered by the hens or by you, forage is essential to provide to your hens year-round. Forage can include weeds, grass, fodder plants, sprouted seeds, compost, excess garden vegetables, kitchen scraps, produce market rejects and discards, silage, insects, mice, and worms. Seaweed, algae, and fungi can also be fed to chickens to supplement their regular feed. The chlorophyll in green leafy plants creates a healthful fat profile and provides hens with plenty of carotenoids that give yolks the deepest orange hues. Along with all those tasty bugs, leafy greens give eggs their maximum farm-fresh flavor. Green plants such as algae or grass pasture will also boost omega-3 fatty acid levels.

Forage is, unfortunately, a missing element in many domesticated chickens' diets. Many undesirable behaviors such as aggression and destructive digging are simply displaced foraging instincts and are uncommon among flocks that are integrated within a garden landscape. Without the ability to forage, hens can become

Opposite, top:
Few things are as satisfying to see as foraging hens.

Opposite, bottom:
If your chickens are confined in an immobile coop, it's especially important to supplement their diets with fresh greens.

IS IT TOXIC?

Many folks are nervous about the possibility of their hens being poisoned by plants while foraging. We respond by agreeing that it is reasonable not to park your mobile coop or pen over plants that are known to be harmful, such as members of the nightshade family, which includes tomatoes. But our foraging hens seem to have no problems avoiding toxic leaves.

Most chickens, particularly the dual-purpose heritage breeds we are most familiar with, have retained from their wild ancestors the ability to detect the bitter flavors of toxic compounds in their forage and avoid potentially poisonous plants and fungi. This has been our experience and that of our customers, although we should caution that this view is contradicted by a handful of apparently knowledgeable members of popular chicken-keeping forums. We accept these firsthand accounts as valid, perhaps indicating regional differences in plant and fungal communities or because of an unseen problem that mimicked a toxic effect. Nevertheless, we recommend that you familiarize yourself with the plant, fungal, and animal community of your area and make an effort not to offer your confined hens anything known to be toxic. Your vet or local agricultural extension service should be able to provide a list of plants in your area known to be toxic to chickens.

Hens know instinctively and with high accuracy which plants are safe to eat and which contain toxins.

obese, can have lower immune function and imbalances of gut microbes, and can develop behavioral problems because they're bored.

For chicken keepers who live in highly urbanized settings, who are dealing with time limitations, or who have mobility constraints, it may be impossible to give hens regular outings to forage. In such situations, if you can manage it, it's worth the effort to do some of the foraging yourself on behalf of your hens. Dig weeds or gather other appropriate fodder (edible plants) in your garden and toss them into the run. Ask a local produce market for wilted greens, or ask a neighbor if he'd be willing to toss freshly dug dandelions into the run. Dandelions are especially good for hens if the roots include a bit of uncontaminated, microbe-rich soil. Just make sure any greens you intend to give your chickens have not been exposed to any kind of herbicide.

By far the most preferable approach, however, is to let your hens forage for themselves in the landscape inside portable day pens or mobile coops. They'll get a greater nutritional boost by selecting their own meals, and they'll gain important benefits from the exercise and being able to engage in enriching natural behaviors.

Our yard consists of three zones that provide for our flock: the vegetable garden, the meadow (a grassy area between trees in our small orchard), and the shady forest edge. We move our mobile coop around the yard into each of these zones by season and need. When we move the coop often, the grazing is light and many nibbled plants survive and regrow. Other times, we choose to leave it in one spot for a while, so that hens can focus their devastating scratching and pecking to clear unwanted vegetation.

MINERAL SUPPLEMENTS

In addition to the macronutrients that hens need in relatively large amounts, they also require a range of micronutrients (vitamins and minerals) in smaller amounts.

Calcium One of the most important supplements to give your laying hens, calcium is the micronutrient used in the greatest quantity, because it's the primary material in eggshells. Some studies of white-egg-laying hens in their first year of production indicate that they require 2.3–2.8 percent calcium in their diets. However, a recent study that examined calcium needs of brown-egg-laying hens in their second year of production (which should correspond better to home flocks), indicate that 3.8 percent is an optimal level for shell quality. Because most compounded feeds contain the lower value, we recommend calcium supplementation for all hens of laying age.

Limestone The most common source of calcium in layer rations, limestone particles come in several sizes, depending on source and grind. Coarse limestone is marketed as combining the functions of grit and limestone. We suspect that a larger particle size limits absorption in the gut substantially, however, making it unsuitable as the sole calcium source. We do recommend it as a supplement and appreciate its versatility.

Oyster shell A rich source of calcium, oyster shell is also valued for its trace mineral content. We like oyster shell as a calcium supplement but not as the sole source of calcium for hens, because of research indicating that it is more absorbable when combined with other types of calcium, such as limestone, than when used alone.

Seaweed You can feed small amounts of fresh or dried seaweed to your hens for a variety of benefits, including calcium and mineral supplementation. Each species of seaweed has its own nutrient levels. We suggest that you follow the feeding advice of the manufacturer of a seaweed product intended for poultry and/or investigate the species present in your area before feeding it to your hens.

Recycled hen eggshells Though eggshells are a popular and free calcium supplement, we recommend against using them in this way because of the risk of encouraging egg-eating behavior. For those who insist on feeding eggshells to hens, we recommend that the shells be baked and coarsely ground in an effort to mask their appearance and taste.

NONTRADITIONAL FEEDS

One of the fun things about chickens is that because they can digest a diverse range of foods, farmers, scientists, and amateurs alike can do a little experimenting with possibilities. These days, people are flocking to a few unusual feeding approaches that hold promise for naturally healthier hens.

Silage By now we hope that you're convinced that foraging on fresh plants is of paramount importance to your hens' well-being. But what about in winter, when much of North America offers little fresh greenery to forage? Those of us with small flocks can borrow a page from farmers who feed their livestock greens in the form of silage in the winter months. Silage is a method of preserving plant material using fermentation rather than dehydration, resulting in winter forage that's more easily digested and loaded with healthful microbes. Making silage is essentially a pickling process that relies on lactic acid produced during anaerobic fermentation to lower pH to hold pathogens and molds at bay.

We have started making silage from our annual abundance of corn stalks and leaves after harvest, though you can use any otherwise edible plant material. Instead of relying on the microbes already present to begin the fermentation, as is traditional, our method stacks the deck by adding bokashi bran, a rich source of desirable lactic acid–producing bacteria.

Sprouted grain fodder We know of some home chicken keepers who sprout whole grains in a similar manner to farmers who feed it to cattle, goats, and other livestock. This handy bit of kitchen counter alchemy transforms raw grains into a form that's easier to digest, neutralizing some antinutritional factors present while providing a rich source of tender greens in any season.

SILAGE

Feed your hens silage as a supplement to their regular diet—a handful every other day for a flock of three will get the job done. Silage can be fed year-round, but hens may lose their taste for it when fresh plants abound.

About 30 pounds coarsely chopped corn stalks and leaves (enough to fill a wheelbarrow)

2 to 3 pounds bokashi bran (or 2 to 3 gallons activated EM/BM)

In the wheelbarrow, blend the bokashi bran (or EM/BM) with the corn stalks. Moisten this with a little water if needed to distribute the bokashi or hydrate the stalks if they are too dry.

Pack the chopped and inoculated stalks into a large, lidded bucket (or several), or a 55-gallon drum with a lid; or simply put them in sturdy garbage bags within a lidded garbage can. Whichever container you choose, your goal is to exclude as much air as possible before sealing it. If you use a container, stuff it full to the brim and mash it down to force out all the air; in a bag, squeeze out all the air, tie it off, and place it in the storage container.

Store the closed containers at temperatures of 60°F to 80°F for three to six weeks (three weeks in warmer conditions, six on the cooler end) to enable the fermentation to work its magic. Then move them to cool area for storage, but avoid freezing, which may make the silage difficult to use.

SPROUTED GRAINS

1 quart (1 liter) water

1 pound whole, uncracked grain

1 teaspoon (5 ml) citric acid

Fill a bucket or large bowl with all ingredients. Set the container aside in a warm area (60°F to 80°F) and let the mixture soak overnight. Then spread the soaked grain into trays at about ⅓ inch thick (thicker could cause them to mold). Porous trays can make draining easier, because you need to rinse the grains two or three times per day with more of the water and citric acid solution, and you don't want the grains to sit in standing water or they will mold. After about a week, the grains will have formed a thick mat that can be easily removed from the tray and tossed to your eager girls.

WATER

Last but not least is a substance that is required for virtually all life on Earth. Although it's well understood by most backyard flock keepers that their chickens require adequate water as well as food, water is usually not appreciated for being the single largest component of their intake and accounting for roughly 85 percent of the weight of a young bird and 70 percent of adult bird. Water is essential for digestion, heat regulation, waste elimination, and various other metabolic functions.

Although most home-raised hens have access to quality city water supplies, variations in salts, pH, mineral content, and temperature may occur. Any or a combination of these can affect flock health and create an invisible factor in disease. If you have doubts about the quality of your water, we recommend having it tested to ensure that no harmful substances are present. If test findings raise too many questions, it's wise to err on the side of caution and consider other options, such as rain barrels, to collect pure water. Once you've tapped a good source, you need to consider how you will be delivering the water to your girls.

Collect the rain off your coop's roof, strain debris, channel the water into a rain barrel, and use it for hen hydration.

Ideally, we would all provide an uninterrupted supply of cool, clean water in an appropriate vessel that is rinsed at each filling and cleaned with a little soap and water and a brush about once a month. Don't worry too much if you're like us and occasionally forget to fill or clean your fount. In normal conditions, your hens can survive a day or two without water, although their laying will suffer, and the risk of dehydration can be very dangerous if temperatures and humidity are high or hens are facing other stresses. The importance of cleaning depends on what kind of waterer you are using. The traditional double-wall metal founts we once used required daily (or twice daily) cleanings to remove debris. The nipple founts we now use require only an occasional rinse.

We're hesitant to say exactly how much water a chicken needs because it varies widely by age, temperature, health, and other factors, but an old rule of thumb is that chickens require two or three parts water to each part of food they consume. If you provide a constant supply of water, the question of quantity may be moot. A flock of five hens can be expected to drink about a gallon a day in warm weather, which is more than our daughters seem to manage despite a combined weight that's twice that of our flock!

Water Founts No matter their form, founts are the drinking vessels chickens use. A standard, doubled-walled galvanized fount, similar in design to traditional feeders, is most commonly used in home flocks. These founts work well and are fairly durable, with a large range of capacities. Being metal, they can be placed directly on a heater base in freezing temperatures to keep the water flowing.

We much prefer a nipple-style fount, however. With this system, ingenious nipple devices are inserted into a water-holding vessel and release small portions of water only when pecked at by a hen. Hens are naturally attracted to the shiny metal tip and learn to use nipples within a few hours. Commercial growers use nipples on elaborate plumbing systems to distribute the water, but home chicken keepers with small flocks can purchase founts with nipples attached to the bottom of hanging 5-gallon buckets. Nipple bucket founts are becoming popular with hobbyists, although their adoption has been far from universal, despite their many advantages.

Rather than purchasing a nipple fount, you can purchase nipples and make your own. Attach a few nipples to the bottom of a bucket or create a PVC pipe setup like ours, attached to a rain barrel by a short length of hose. The only real drawback

Our flock shares clean water from a nipple bucket fount. A homemade nipple waterer we created with PVC is on the right.

RECOMMENDED WATER ADDITIONS

Aloe The juice of *Aloe barbadensis* has been shown in research to be valuable in protecting hens from a range of maladies of the digestive system. We especially value its ability to protect against coccidiosis, a serious disease in chicks, which helps you avoid the use of pharmaceuticals.

Apple cider vinegar In small amounts, apple cider vinegar could acidify the water just enough to suppress proliferation of disease-causing microbes such as salmonella and eimeria (the protozoa responsible for coccidiosis). At higher concentrations, acidic water will erode the galvanized zinc coating on metal founts, releasing it into the water and exposing the metal below, which will rust. Avoid using apple cider vinegar in very hot weather when you're using alkalizing additives such as baking soda, because the vinegar will counteract the alkalizing effects.

Effective Microorganisms (EM) or other beneficial microbes (BMs) Studies show that providing EM or similar beneficial microbe formulations improves the availability of nutrients, helps equilibrate the intestinal microflora, reduces incidences of diarrhea, improves egg formation (because of better assimilation of calcium), and improves egg flavor and yolk color.

Garlic Known as an excellent overall supplement and immune booster, garlic can be added to water in very small amounts at first, to help hens get used to the taste. We've gradually built up to about one crushed clove dropped in our 5-gallon fount, which we replace at each filling. There are various commercial options out there, but we prefer to use the fresh stuff, mainly because we grow garlic ourselves and we always have a few unsightly cloves around to give to the hens.

Humic acid This organic substance results from the decomposition of organic matter—particularly plants. When included in water and feed, humic acid has been shown to inhibit bacterial and fungal growth and provide antiviral and anti-inflammatory properties; it also improves hens' immune systems, helps reduce stress, and prevents and cures intestinal disorders.

Thyme essential oil This is one of the most well-studied and proven effective preventative treatments for a variety of conditions that affect poultry. It's wise to start your chicks or hens on a tiny amount (dip a toothpick into the oil and swirl it in the water) to get them used to the strong flavor—they can resent it at higher levels. Gradually build to one or two drops per gallon with each water change.

of this system is that it's impossible to put additives in the water without having them become diluted by incoming rainwater. In the future, we'll try adding a reservoir between the PVC nipples and the barrel, but for now, the PVC fount makes a terrific backup on hot days, and the barrel makes a convenient tap from which to fill buckets.

One drawback of nipples is that they will eventually freeze up in cold weather. We solved this problem by bringing in the hanging nipple buckets each night that we expect a hard freeze, placing them inside regular buckets to prevent messy drips. In the morning, we return the founts to their spots in the run, where they provide water long enough to keep hens hydrated for the day before freezing again.

Traditional feeders are wasteful and invite mold and rodent problems.

Above right: Commercial treadle feeders can be pretty expensive, but you can find some reasonably priced options online or build your own if you're handy.

Nipples will occasionally break if handled roughly by hens, and, depending on your setup, broken nipples can be difficult to remove and replace. After removing a broken nipple, you can plug up the hole from inside the bucket using swimmer's earplug wax. Not a swimmer? Try chewing gum, softened beeswax, or a similar nontoxic material.

To prevent leaks and raise the water out of the reach of thirsty rodents, nipple waterers must hang more or less vertically, usually over the hens' heads. Hanging our nipple bucket from the top of the run using bungee cords works well for us with one drawback: our hens' drinking gradually reduces the weight of the water in the bucket, causing the bungee to retract, in turn raising the bucket until it's hanging nearly out of their reach. We solved the problem by adding a few large chunks of wood below the bucket to provide a sort of step stool for them. This also helps small, newly introduced pullets to reach the fount.

FEEDING EQUIPMENT

The standard feeder for many home flocks is the hanging, 12-pound galvanized feeder. Its large capacity is a virtue when you leave home for vacation, and filling it once a week is an attractive prospect, but like many other good things, it's too good to be true. The feeder's size is ultimately its undoing, offering far more feed than most small flocks can consume before it becomes stale or spoils. And it serves up an all-you-can-eat rodent buffet.

Our favorites, treadle feeders are commonly used in the United Kingdom, but they are not as widely used in the United States. These ingenious feeders are essentially covered hoppers full of food that let down easily into a covered trough. The lid on the trough can be lifted only by the weight of one or more hens depressing

FEED QUALITY

Our current store is housed in a former brass foundry in the central part of the city of Portland, on the east side, in an area cleverly called the Central Eastside. This district is situated directly across the Willamette River from downtown, and it's long been the place to go if you need parts cast, foam rubber cut to shape, a transmission rebuilt, or industrial grinding (whatever that is). But all that is about to change. As we write, four low-rise, mixed-use condos or apartment buildings are being built around us. Directly across the street, two blocks of old warehouses are being carved into a "working hub for the regional food economy" by a local nonprofit. And, this being Portland, our neighbors include no fewer than three microbreweries, two distilleries, two marijuana dispensaries, one bikini coffee cart, and one pingpong–themed bar. And then there's us, a chicken and garden store, perhaps the most surreally Portland stereotype of them all.

At any given time, ten to fifteen pallets stacked with feed bags are parked on our dusty and uneven concrete floor. Some hold bags of plain cracked grains, but most hold large brown or white bags full of pelletized feed—an unusually wide selection of complete diets for laying hens. Over by the chick pens are smaller bags, plump and floppy like piles of white sausages, filled with finely ground starter feed packed with enough extra protein to grow a chick into a hen in a relatively few weeks.

We stock these many feeds and are so passionate about the subject for one reason: The quality of nutrition you provide your hens affects every aspect of their lives, from their overall health and longevity to egg production and temperament. It naturally follows that you should provide the highest quality food for your flock that you can afford, and, beyond this, you should modify this diet to respond to seasonal demands, available resources, and your hens' individual needs. When you consider that 65 percent of a hen's immune system resides in her gut, the link between health and diet comes into even sharper focus.

Furthermore, feed varies widely in the manner in which it's grown, processed, and transported. Most of our customers would agree that urban chicken keeping, while not a political act, carries a general ethos of respect for the land and self-sufficiency that demands that we make sustainable selections when possible. At our store, we offer everything from farm chow layer rations for those on a tight budget, to organic, regionally grown, vegan, artisanal blends crafted by young men with large beards in solar-powered workshops. Okay, the last part is a joke—the men aren't actually young. There are many feeding decisions you can make, and your choices are important on many levels. For some of our patrons, environmental and sustainability issues dominate their feed decisions. Others seek out feed and supplements that will enhance their hens' eggs with healthy fats. Those of us who've seen a few too many chick seasons are usually content to find feed in a bag small enough to carry in from the car without straining an aging back.

HOW MUCH FEED?

If you're wondering how much feed to buy, consider that a flock of three laying hens will consume about 50 pounds of feed in six weeks. We feed our current flock of six hens from 40-pound bags of feed, which works out to a fresh bag every eighteen days or so. If we didn't own a feed store, we'd probably buy two bags at a time every month, give or take a couple of days.

a platform moved by levers, thus keeping the food protected and dry until a hen moves into position to eat. This keeps rodents and moisture on the outside, where they belong.

If a treadle feeder seems too expensive, you can use a simple poultry-sized feeding trough, adding measured scoops of feed for the day. This will keep the feed fresh and dry, while avoiding excesses, but it does require daily replenishing.

FEED STORAGE

A galvanized garbage can with a tight-fitting lid provides water-resistant, rodent-proof feed storage.

Chicken feed has a shelf life of about six months when it's stored in cool, dry conditions such as in a basement (or a feed store). To protect your precious feed from meeting an inglorious fate, its proper storage is of the utmost importance. We recommend that you store feed for the long term indoors in a chew-resistant container that's as airtight as possible and in a cool, dark place. When feed is stored for shorter periods of time, conditions need not be so exact.

We also recommend that you not place feed bags directly on the floor, because moisture may move up into the bag, hastening spoilage and mold formation. Even bigger risks of informal storage are pests: nutrient-rich chicken food is extremely attractive to critters such as insects, rats, mice, and raccoons.

At home, we store about a month's worth of feed outside near the coop in a metal garbage can with a tight-fitting lid. We've not yet been bothered by pests or mold problems within the container as long as we use all the feed within a month. If you opt to store feed in this way, and raccoons or other marauders are known in your area, you can secure the lid tightly with rope or bungees.

MOLD AND MYCOTOXINS

Feed arrives fresh from the mill containing 7–10 percent moisture. Feed will absorb moisture from the air at an increasing rate above 65 percent humidity, and mold spores will sprout and grow within the feed after its internal moisture rises above about 15 percent. Once mold is in your feeder, even fresh feed can become stale or moldy in a matter of days, particularly in humid conditions.

Molds produce mycotoxins that, when eaten in small doses over long periods of time, can cause a chronic reduction in your hens' overall health—and in higher concentrations, it can cause illness or death. Although mycotoxins are found in nature, chickens encounter them primarily in feed. We suspect that these toxins are the underlying cause of many seemingly mysterious poultry problems, ranging from reduced laying, to illnesses that are difficult to diagnose, to mortality.

All feeds contain small amounts of these toxins. Although there's little that you can do at home to prevent contamination that may have occurred in a batch of grain at the mill, you can take some comfort in knowing that mills do test samples regularly and will probably catch a bad batch before it gets to you.

COMMON FEEDING PROBLEMS AND SOLUTIONS

Beyond its direct effect on hen health, feed is often at the root of many seemingly unrelated problems. We've listed the top five commonly encountered feed- and water-related chicken-keeping problems and our recommended solutions.

Problem: Overfeeding sugars and carbs We've witnesses an epidemic of chicken keepers loving their hens into metabolic problems by giving them too many treats that are high in carbohydrates and/or sugar. Because hens can eat only a limited volume of food and will always choose to eat carbs first, the net effect is a mild to severe dilution of the food value of their primary, nutritionally balanced feed. It's truly difficult to acknowledge the seriousness of this situation, because the effects are not sudden, but incremental, and they manifest in unexpected ways. Overfed obese hens can experience egg binding, heart attacks, and increased sensitivity to stresses such as extremes of heat and cold. These hens may suffer from an altered microbial community that underlies various diseases of the digestive tract. And these protein-deficient hens produce smaller and fewer eggs and sometimes engage in cannibalism, feather picking, aggressive behavior, and/or egg eating.

Grains and other carbohydrates should be a sometimes treat.

Recommended solutions Gradually and steadily reduce the amount of high-carb treats and scraps you give your hens. Start by cutting back by a third the first week, and the next week cut another third. Unless you are grossly overfeeding carbs to your hens, this should be enough to get you back in the appropriate range. During hot weather, eliminate all extra carbs, with the possible exception of small amounts of frozen fruit to keep your hens' metabolisms from over-revving. Reverse this in winter: Hens need more sugars for warmth, but don't overdo it. Never feed more than a handful of carb-rich snacks per day for every three chickens in your flock.

Although hens can develop carbohydrate addictions that can be hazardous to their health, you should provide a moderate amount of grains in a variety of particle sizes, which aid digestion while improving gizzard tone. We have switched to feeding our hens sunflower seeds as treats in summer because of the protein they offer, along with larger particles and fiber. In winter, we feed them a couple of handfuls of cracked corn on cold days; otherwise, we don't toss in much other than dandelions and other tender perennial weeds, which they really appreciate.

Many of our customers have switched to using dehydrated mealworms for treats, which most hens actually prefer to corn. They are very rich in protein (about 50 percent) and fat (about 25 percent). As with most treats, avoid overfeeding mealworms to your hens to avoid problems resulting from excess protein.

Problem: Lack of portion control The standard feeder for home flocks is a cylindrical, galvanized metal contraption that's hung within reach of your hens. Most hold a whopping 12 pounds of food (some are bigger yet), which doesn't sound like much to folks accustomed to buying 40- or 50-pound bags of feed. Consider, though, that each hen consumes only 4 to 6 ounces of feed daily, and even less if she consumes scraps and forage. A 12-pound feeder holds enough feed for about ten days for a flock of three—if rats and mice don't find it first (they will), or it doesn't go stale (it does), or it doesn't become too moldy to eat (in damp conditions). These feeders also tend to have problems letting down their contents and must be shaken daily to release the stuck food. No fun—and not good for your hens.

Recommended solutions The simplest solution is to provide only enough feed to the hens for the day, and store the feed you will not give them that day in a secure storage bin or trash can. It's fine to fill up the feeder with a few days' worth before you head out for a trip, but on all other days, source food from the bin in daily amounts. To use this strategy, you'll first need to figure out how much your flock is consuming and place only that much in the feeder each morning. Get into the habit of feeding when you visit the hens to greet them, inspect them, collect eggs, and clean the area if you are one of those daily cleaners who are rumored to exist. The process is straightforward: Include 1 cup (8 ounces) of feed per bird into the feeder in the morning and return at night to measure how much is left. Subtract this much from the amount you started with, and voilà! You now know how much they are able to eat in one day. If you want to be very accurate, repeat this for a few days and average your results. Keep in mind any other sources of nutrition (such as foraged greens). If you are measuring during the summer, repeat this measurement exercise during the winter, when they usually eat more.

Problem: Rats and mice Any food that is outside, including vegetables growing a garden, trash in a trash can, or compost, will attract small, hungry animals with

a keen sense of smell. Chicken food is especially attractive to rats and is, in fact, similar in composition to pet rat food. Mice are usually less of a problem because chickens see them as prey if they can catch them, whereas a rat is too large for a hen to handle (roosters will take them on, however).

Recommended solutions Your first instinct may be to try to build a rat-proof fortress. This may involve the use of hardware cloth, which is just about the most expensive kind of wire fencing you can buy. It's been our experience, however, that if food is present and constantly available, rats will eventually find a way to get in. Instead of spending money trying to build your way out of rodent problems, we suggest that you purchase higher quality feed and a treadle feeder.

Problem: Feed is too expensive We've calculated that feeding a typical backyard flock (three hens) costs between fifty cents and a dollar daily, depending on your location, feed selection, and supplementation routine. Assuming that you didn't spend a large sum building your coop and buying supplies, this is a fairly modest investment for the fresh eggs you receive in return. If other important benefits of chicken ownership, such as their value as pets—priceless!—are factored in, the bargain gets better. Still, this represents an expenditure of about U.S. $400 a year if you choose a premium, organic layer food plus a few supplements and treats. If you get hooked on chickens like we did, we can assure you that three hens will not be enough (they're more addicting than potato chips!) and you will be paying proportionately more to feed all those hungry beaks.

Recommended solutions A proper feed ration is an essential base of balanced nutrition for layers, and we believe they need to eat some pelleted food to be optimally productive and healthy. You can stretch your feed dollar by allowing your hens to forage for some of their food in your yard or garden. This alone can shave 10 percent or more off their feed consumption when plants are actively growing and bugs are abundant—but if you live anywhere but the tropics, seasonal variation will diminish their availability significantly.

Another strategy is to make your own versions of feed, scratch grains, and supplements. We offer plenty of ideas in this book, but realize that DIY feeds can be difficult to get right, requiring exacting proportions of ingredients. To save money by making your own feed, you'll need to be buying (and storing) 50-pound sacks of feed ingredients or growing and storing your own. If you have a small flock or lack the time to devote to such a project, you're likely better off skipping this process and seeking savings elsewhere.

Chicory, *Cichorium intybus*, is an ever-present, nutritious weed.

Plantain, *Plantago major*, seems to prefer to grow where other plants cannot: in wheel ruts, cracks in concrete, and in the poorest soil possible. It is delectable to chickens.

Dandelion, *Taraxacum officinale*, is a gardener's least favorite weed but a fantastic chicken treat. More motivation to dig them out!

Problem: Lack of green forage We keep returning to this message, but it's worth repeating that hens have been bred to thrive on a mixed diet of both formulated feed and food obtained from foraging. Forage provides essential chlorophyll, enzymes, animal protein, and a raft of beneficial microbes that complement formulated feeds.

Recommended solutions If your hens cannot free-range forage for themselves and you cannot make use of a mobile coop, commit to patrolling your garden regularly with a hoe in hand to gather weeds to feed to your hens. Start with common and familiar weeds such as chicory, dandelion, and plantain, which are all safe for your hens to eat—if they are not contaminated in some way, such as by pesticide use. You may also find or grow many more plants that your hens can safely consume.

WHOLE HEN HEALTH

MAINTAINING A HEALTHY FLOCK

Quality veterinary care for chickens is sometimes difficult to find and always expensive. It's for this reason that the most requested thing at the Urban Farm Store is neither a feed nor a supply, but chicken health advice. As you might expect, given our lack of formal veterinary training and certification, our prescriptions focus primarily on prevention using the tools we outline in this book, such as dietary improvements and environmental modifications. But when the time for prevention has passed and a customer has a sick or injured chicken at home, we try to guide her through simple procedures or suggest an obtainable remedy that may help.

Chickens are unfamiliar to most urban and suburban veterinarians, because even vets who treat birds are accustomed to the cage-dwelling types that are not exposed to the elements or many outdoor pathogens. With the ever-growing popularity of small-scale chicken keeping, however, city and suburban vets are becoming more adept at treating chickens. Farm vets may know a bit more about chickens, but they are likely to take the livestock husbandry approach to health issues in a flock—with a probable recommendation to cull (kill) individuals or even whole flocks.

For a family that keeps chickens as beloved pets that happen to lay eggs, the idea of killing one for the good of the rest of the flock is unimaginable. Because each family, or even each member of the same family, has a different feeling about where their hens lie on the pet-to-livestock continuum, decisions about providing chicken healthcare can be more complicated, both emotionally and financially, than making similar decisions for a pet cat or dog. It also has serious implications regarding what to do with a hen that is done laying or a chick that turns out to be a rooster.

On most farms, chickens are viewed as livestock. This usually means the expense of medically treating an individual bird is not generally worth the dollar value of replacing it. A bird with a life-threatening wound, for example, would be euthanized or left untreated to let nature take its course. On the other hand, when an illness could affect an entire flock, all the birds are generally treated at once, often

A healthy hen looks perky, with bright eyes and good-looking feathers, and she moves with vigor.

with antibiotics added to the watering system, to prevent the spread of the disease. Happily, many common chicken ailments are treatable without antibiotics or surgery. Of course, you won't be able to solve every issue without the help of a professional, but you can treat most.

Although it's nice to have skilled vets available for backup, our goal is to teach you strategies to avoid the vet for all but the most dire cases. Prevention is truly the best strategy, and it is much less expensive as well. Maintaining a healthy and safe environment for your chickens is absolutely essential to the long-term wellbeing of your flock and will prevent most of the health issues we discuss. If your chickens live in an environment that is predator-proof and they are fed high-quality food and supplements, they are likely to remain healthy. Most chicken injuries are the result of predator attacks or coop building flaws. Most diseases are either brought in with chickens from other flocks or result from environmental issues. A well-cared-for chicken will have a robust immune system that can withstand the occasional stressor such as illness or injury.

If you want to understand more about the science behind chicken healthcare in flocks in earnest, we recommend Gail Damerow's very thorough and informative book, *The Chicken Health Handbook: A Complete Guide to Maximizing Flock Health and Dealing with Disease*. Her expertise can guide you through such complex endeavors as collecting and analyzing chicken stool samples and differentiating between coccidiosis strains.

CHICKEN LITTLE VERSUS BIG PHARMA

By the mid-twentieth century, traditional chicken-keeping methods had largely given way to modern intensive feeding and confinement practices and mega-farms. Even as productivity soared, it soon became apparent that high population densities and artificial environments were stressful to the animals. Not only were these conditions cruel, they sometimes resulted in sweeping losses caused by highly transmittable diseases. Rather than address the root causes, farmers increasingly turned to antibiotics for both disease treatment and prevention. When it was observed that chickens receiving these treatments grew faster than their "healthy" counterparts, antibiotics found a seemingly permanent place on the menu as a growth stimulator. The alarming result was that these drugs, important both to livestock and people, became less effective or ineffective as pathogens gradually adapted to them.

European regulators were the first to confront the menace of antibiotic-resistant super bugs, banning antibiotics in feed in 2006. The United States was slower to react but in 2015 the FDA enacted the Veterinary Feed Directive. We learned of the VFD shortly before enforcement began in 2017 and were shocked to learn that hobbyist chicken keepers would be regulated identically to famers, underscoring the unique dual-nature of chickens as both pets and farm animals.

We seldom gave our hens antibiotics, and we never used them to increase growth or cover for poor conditions, but we were legitimately concerned about their loss as an important bulwark against a handful of potentially fatal poultry diseases. Our initial shock soon gave way to enthusiasm, however, as it dawned on us that the VFD could help usher in a new, more compassionate and natural approach to poultry care both in the farmyard and in the backyard.

THE BODY OF KNOWLEDGE

Becoming an active partner in enhancing your flock's health begins with your understanding of the basics of chicken physiology. If that flashes you back to high school biology class, don't fret—there are just a few key things to know, and we won't quiz you or make you dissect a chicken! We simply want to establish a common vocabulary and describe how a healthy hen looks and acts and what can be observed that may indicate a problem. Lacking a vet's ability to run lab tests, these outward clues are especially important to small flock keepers, because they provide our only glimpse of our hens' inner condition.

Becoming familiar with a chicken's physiology is your first step toward being able to diagnose health problems.

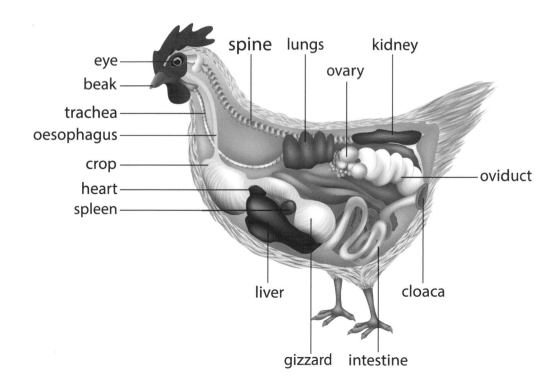

eye
beak
trachea
oesophagus
crop
heart
spleen
spine
lungs
ovary
kidney
oviduct
liver
gizzard
intestine
cloaca

PERFORMING A BASIC CHICKEN EXAM

A simple exam can provide valuable information that may lead to a diagnosis.

The best way to become acquainted with your chicken's body is to pick her up and practice a top-down examination. Go ahead, try grabbing a chicken. What do you notice?

Hard to catch	Good sign! She's frisky!
Looks frisky, allows you to pick her up	She's a tame, well socialized hen.
Moves slowly if at all; looks droopy, oily, and/or fluffed-up; is easily grabbed	She's lethargic, a serious symptom of several health conditions, some life-threatening.
Lighter than expected, particularly when compared to her flockmates	She's not eating properly, is dehydrated, or suffers from worms or other internal parasites—a highly concerning signal.
Heavier than expected	She's obese, often resulting from excessive carb-rich treats; or she could have a bowel or crop obstruction.
Aggressive, pecks hand, growls, remains on nest	She's a broody hen defending her eggs.
Meows and rubs leg	That's not a chicken! It's your cat!

Examine the plumage on her back, belly, tail, and wings. Part the feathers to examine the shafts, points of attachment, and the skin they conceal.

Feathers glossy, slightly oily, clean, and undamaged	She's healthy.
Skin white, yellow, or black	Could be normal, as skin color varies by breed.
Feathers appear chewed at the base, skin irritated and scabby	She may have mites (look for tiny red or black specs near the bases of feathers).
Reddened skin around vent, clusters of eggs on base of feathers, visible insects	She's got lice!
Patchy feathers or dry bald spots on back, neck, or vent	She's been picked on by her flockmates, or may be molting.
Feathers absent around wings and tail	She may be pecking herself because of heat stress or parasites.
Bare skin on chest	She may have breast blister, or it could be wear nd tear from rubbing against a fence. This is often found in cases of pendulous crop.

Next, examine her legs and feet.

Skin yellow, gray, or black with smooth, shiny scales that lie flat	This is normal (younger hens have more pigment than older ones).
Scales raised, crusty, irregular	She has scaly leg mite (may be harder to see and treat on feather-legged breeds).
Feet swollen and/or large sore on pad	She has an infection, possibly bumblefoot.

Continue your exam by checking the condition of her comb and wattles. Size and shape will vary by sex and breed, ranging from compact and bumpy to large, smooth, and even a bit floppy.

Deep, even red color	This is normal for a mature hen during laying season.
Light shade of red, smaller than usual but not shriveled	This is normal for a hen who is not laying (young or off season).
Bright red and swollen	She could have an injury, infection, or fever.
Pale pink, shriveled	She may have anemia caused by red mites (which are visible only at night).
White flakes on one side only	She may have a fungal infection.
Black or black spots	In winter, she could have frostbite; in warmer months, it could be dried blood. Also consider advanced fowl pox.

White spots associated with blisters and scabs	She likely has fowl pox.
Scales on face, comb, or wattles	Look for other evidence of northern fowl mite.
Comb on side of head in shades of white, red, or blue	Those are her earlobes, and they correspond to egg color!

A chicken's eyes are set on the sides of her head, so she must turn to one side or another to get a good look at something of interest. Of interest to us are the many things her eyes can tell is about her health.

Clear lenses, round shiny pupils, no discharge	Her eyes are healthy.
Iris is brown, black, gold, red, or pink (albino)	This is a normal range of coloration.
Pupil or iris irregularly shaped	She may have nerve tumors, likely Marek's disease.
Raised, dark, blisterlike lesions on skin around eye	She may have avian pox, a viral condition. Lesions can also occur on other areas of skin not covered by feathers such as legs, wattle, comb, or vent.
Red, swollen eyelids; avoids light	She has conjunctivitis, often caused by highly pungent ammonia gas from bacteria in damp litter saturated with feces.
Eyeball or lids swollen, pus present	She may have a bacterial infection, possibly salmonella.

Yellow plaques under eyelid, inflammation	She may have been exposed to aspergillus or another mold, typically from damp litter or contaminated feed. May also affect respiratory system.
Cataracts	She has a vitamin E deficiency.

While you're examining features on the head, turn your attention to the nostrils and beak while listening for breathing issues.

Nostrils (nares) clear, breathing regular, interior of mouth and tongue smooth	This is normal.
Beak smooth, hard, top and bottom fit	This is normal.
Watery discharge from nostrils (and eyes), coughing or sneezing	She probably has nonspecific respiratory disease.
Mucous discharge from nostrils, facial swelling, tearing or bubbles in corner of eyes	She has a serious respiratory illness that could threaten the lives of the entire flock.
Cankerlike lesions in mouth	She has fowl pox, wet form, or avian trichomoniasis.
Beak offset or twisted looking	She has crossbeak.
Tip of beak abnormally long, sharp, and curved	She has parrot beak. Her beak needs a light trim to allow her to eat normally.
Neck stretched out, mouth open, gasping for air	She has gapeworm.

Finally, take a good look and gentle feel of her undercarriage, including her breast, abdomen, and vent area below her tail.

Breast soft, may be puffed out with grainy material, palpable	This is normal (within breast is the crop, the short-term food holding organ).
Abdomen not hard, bony ridge of keel bone palpable under some fat	This is desirable and normal.
Breast or crop enlarged, hard	She has an impacted crop (may let down, so reexamine in a few hours).
Crop enlarged, hanging low, often missing feathers on chest	She has a pendulous crop.
Crop enlarged, fluid filled; gentle squeeze produces foul liquid	She has sour crop.
Blisterlike bubble of fluid on keel bone	She has breast blister.
Abdomen swollen	She may have heart disease, tumors, impacted eggs, or cysts in her oviducts.
White discharge from vent, feathers missing, redness with yeasty smell	She has vent gleet.

EPSOM SALT BATH

On a feed delivery not long ago, Robert was astonished to meet a small flock of hens that were robust and laying regularly at the ripe old age of eight years, nearly double the productive life of most chickens. When asked about his secret to success, the chicken keeper was at first coy, offering only that the flock liked to roam his garden and was lucky to have avoided becoming dinner for predators. Soon he revealed his secret: Epsom salt soaks. If one of his hens looked sluggish or dull, he placed her in a small tub of prepared water and gently bathed her, a relaxing and rejuvenating experience for his girls he dubbed "the chicken spa."

An Epsom salt bath has become a go-to solution for us as well, and you can do it, too. Fill a chicken-sized basin with 2 or 3 gallons of warm (not hot) water and ¼ cup Epsom salt. Set this up somewhere warm and separate from the other chickens. A bathroom is the obvious choice, and putting the basin in your bathtub will help contain your little lady if she gets frisky or splashes out a lot of water.

Capture your hen and place her gently in the bath. Believe it or not, most chickens love baths. If she needs

Roxy is calm but perhaps a bit perplexed during her first Epsom salt bath.

calming, however, carefully drape a towel over her. Let her soak until the water cools. If her bottom needs washing, the time to do it is toward the end of the bath.

When she's done soaking, wrap her snugly in a towel. Then turn your blow-dryer on low and dry her off. Despite the loud dryer noise, most of our chickens adore this treatment.

FLOCK FIXERS »

Chickens are hardy creatures with a remarkable ability to heal. As their caretakers, we need to be ready when they need an assist, as is often the case when they are sick or injured. In addition to the products we use to maintain health, Hannah keeps a variety of items near the coop for emergencies and the treatment of illnesses should they arise.

You may notice that several items on the list would not be considered organic. We have included them because of their utility, and we don't hesitate to recommend them, especially for emergency use or for your hens' long-term health. Our standards are high, and you can be sure that only preparations with ingredients that are proven to be safe and beneficial for the intended application have been included. Most of these are store-bought items that we've used or are familiar with.

Some of the items we recommend keeping on hand in a basic first-aid kit

First aid for wounds, breaks, and topical care

✦ **Colloidal silver** Antimicrobial liquid for topical or nasal use only.

✦ **Disposable gloves** For managing anything gross or septic.

✦ **Dr. Naylor Blu-Kote** Antiseptic, antifungal liquid that dyes wounds and rashes blue to deter flockmates from wound pecking.

✦ **Duct tape** Water-resistant covering for keeping wound dressings in place. Good over gauze to hold splits when setting broken bones. Use dark tape to discourage pecking.

✦ **Finger splints** Perfect for setting broken legs (as Hannah did this summer).

✦ **Gauze** 2-inch-by-2-inch pads, and 2-inch stretch gauze for wound dressings.

✦ **Gentian violet** Traditional and effective antifungal medication for internal and external application.

+ **GreenGoo Animal First Aid** Organic salve for wound treatment.

+ **Iodine** For cleansing wounds if you don't mind an orange chicken (and don't have Vetericyn on hand).

+ **Manuka or honeydew honey** Organic and effective alternative for topical wound treatment and prevention of infection.

+ **Preparation H** Vasoconstrictor that does an excellent job of shrinking swollen tissues.

+ **Syringes (disposable) with 12-gauge, blunt-tipped needles** For administering oral fluids and for emergency treatment of an egg-bound hen; they don't need to be sterile.

+ **Vetericyn Plus Poultry Care** Stabilized hypochlorous spray solution (not the same thing as bleach) with very good evidence for effectiveness as a wound wash; it's pH balanced, so it's nonirritating and safe even for use in eyes. Our go-to spray for major injuries.

First aid for illness and immune support

+ **Apple cider vinegar** Popular as a water acidifier, general tonic, and source of probiotics. Select a raw, unpasteurized brand with live cultures if possible, or make your own.

+ **Big Ole Bird Poultry Supplement** Organic liquid probiotic and mineral formula that we use regularly in our flock's drinking water.

+ **DuraStat with Oregano** Water-soluble supplement containing oregano, rosemary, sage, and cinnamon that's capable of much more than "stimulating water intake" as the label suggests. Contains well-researched botanical compounds proven effective against a variety of common pathogens and the diseases they cause.

+ **Epsom salt** For chicken bathing and/or gastrointestinal flushes.

+ **Rooster Booster Vitamins & Electrolytes with Lactobacillus** Water-soluble probiotic, vitamin, and electrolyte powder.

+ **VetRx Poultry Remedy** Botanical-based remedy recommendable for minor chicken respiratory ailments. Also said to be effective against leg mites.

TREATING HENS WITH BOTANICALS

As passionate gardeners, we have long enjoyed growing and harvesting a variety of herbs for cooking or sipping as tea in winter. Sit down with Robert for a steaming mug, and he'll rattle off the traditional uses and curative powers of the herbs you are enjoying. Hannah, daughter of an immunological research scientist, would like to see the meta-analysis of double-blind, clinical trials published in a respected journal before she'll have a sip. So naturally, Hannah was assigned the task of writing about the current state of the art, or more accurately science, of treating hens with botanicals.

After surveying and reading several dozen published studies substantiating the value of botanical and other natural healing substances for use on poultry, Hannah is now convinced that certain herbs can be used effectively and safely for chickens. She combed the scientific literature to put together a list of beneficial remedies with summaries of how you can use them. She also includes a few remedies that are often recommended elsewhere, but that she would not recommend based on the research she found. All of these botanicals can be purchased from botanical shops or online retailers.

AGARICUS MUSHROOM

An extract made from *Agaricus blazei* Murill has been shown to offer protective immune system and antioxidant benefits in numerous published studies. In an unpublished study, Atlas, the company that markets the Agaricus Bio product for chickens, suggests that adding it to poultry water increased nutrients and decreased cholesterol in eggs, produced denser egg whites, and increased egg production in older hens. To obtain similar benefits, according to the study, mix five to ten drops of Agaricus Bio per 1 gallon of the flock's drinking water. We do not recommend home cultivation of this mushroom unless you are a mushroom-growing expert.

ALOE

Research in poultry has shown significant benefit in whole leaf *Aloe barbadensis* juice supplementation against coccidiosis. We recommend a continual dosing of ¼ cup fresh aloe juice per quart of water, or ¼ teaspoon of concentrated powder per quart. *Aloe ferox* is available in powder form and has been shown to be highly effective against parasitic worms, at doses of ⅔ teaspoon per gallon of drinking water. This aloe is also effective at treating Newcastle disease and salmonella infections. These doses are safe for long-term use with chickens. Because *A. barbadensis* and *A. ferox* share many constituents, we generally just use *A. barbadensis*, having come to the conclusion that the benefits shown for one form of aloe are likely to apply to the other.

ARTEMISIA, OR SWEET WORMWOOD

Artemisia annua is an amazing herb. Because of its safety and remarkable benefits, we recommend using it daily to supplement your hens' food. A valuable alternative to synthetic medications for coccidiosis, artemisia powder was used in studies to treat infected chickens and they fared better than those treated with a synthetic coccidiostat. In addition, *E. coli* and other coliform bacteria populations in chicken intestines were shown to be lower when the birds were fed *A. annua* leaf powder at 1 tablespoon per quart of feed. Even at high doses provided as feed replacement (5 percent of the diet), artemisia leaf was safe and showed a marked effect in providing antioxidant support and increasing gut acidity. Its extract has been shown to speed wound healing when applied topically, with similar results to the commonly used hospital antimicrobial silver sulfadiazine. Easy to grow from seed, *A. annua* grows high and lanky, which makes it messy as a garden plant but easy to harvest and dry in bunches. It does self-sow, so deadhead or gather it all before it goes to seed if you don't want to see it again next year. The ornamental artemisia, *A. absinthum*, is a lovely perennial that is common in ornamental gardens, but little data is available to prove its safety, and it has lower levels of artemisian, the compound that appears to provide much of the health benefit.

BLACK CUMIN

Studies have suggested that black cumin (*Nigella sativa*) supplementation decreased egg cholesterol levels and increased egg production, egg weight, and shell thickness. Another study showed that it also decreases *E. coli* colonization while increasing beneficial bacteria in hens' digestive tracts. Other studies demonstrated that black cumin enhanced immunity against the Newcastle disease virus. It has also been shown to decrease anemia, and it helps improve chances of survival from illness caused by environmental toxins. To obtain these benefits, mix 1 cup of black cumin seed meal with 3 pounds of chicken feed.

CLOVE OIL

Clove (*Eugenia caryophyllus*) oil studies show that it works particularly well against insect pests when it's sprayed in high-humidity or high-dust environments, so we include a few drops of it in our homemade coop spray. Use caution when handling clove oil and don't spray it directly on your chickens—in its concentrated form, it is a serious skin irritant. About 90 percent of clove oil consists of eugenol, a compound that works better than chlorine to kill staphylococcus on eggs. However, we've found that it doesn't work quite as well as oregano, plus it imparts a distinct odor to the eggs. Clove is a common spice that's readily available as an essential oil.

FEVERFEW

Feverfew (*Tanacetum parthenium*) is often recommended as an insect repellant in chicken coops because the herb is traditionally used in human herbal medicine and has some similarities with *T. cinerariifolium*, from which pyrethrin (an effective insecticide) is derived. However, the only study we found regarding feverfew use in poultry

LEARNING (AND SOMETIMES FORGETTING) *our* LESSONS

Virtually every time we've accepted new chickens, we've regretted it unless we've followed our own advice and quarantined the new birds. Setting up two separate coops makes this possible. We keep the new chickens in the extra coop for a few months while we observe them and make sure they show no evidence of illness. If in this time we observe any signs of illness, such as sneezing or runny chicken noses, we can assess whether they are otherwise fine (probably just nonspecific respiratory disease) or something more serious is in play. Sometimes this quarantine also gives us the opportunity to diagnose some less serious but still irksome ailment such as a mite or lice problem, so we can treat and cure any new birds before causing a flock-wide infestation. But sometimes we're forced to stop and reconsider the wisdom of our adoption plan—or lack of it.

One midsummer day, while slogging through the final edits for this book, we noticed that several hens in our main (stationary) coop were starting to look a bit unkempt. Hannah, curious about their condition, picked up a nearby hen, parted her belly feathers, and was horrified by what she found: dozens of plump poultry lice scurried over irritated skin, feeding and tending to nauseatingly large clusters of tiny eggs.

We concluded that the trouble began about three months earlier when, harried and overconfident after many lice- and mite-free years, we added a hen from another flock to ours without a thought to basic quarantine procedures. The lice must have jumped from the new hen onto the others, and now we faced a crisis serious enough to affect the health and laying of otherwise healthy hens.

Our first instinct was to dust the girls with the relatively mild, food-grade diatomaceous earth powder, but because the extent of the problem was so alarming, we opted for standard poultry dust containing pyrethrum, a moderately potent insecticide synthesized from a natural source. We applied the dust to the entire flock, repeating the treatment three times, seven to twelve days apart. Though there was noticeable improvement, the dust did not provide total elimination, perhaps in part because of our irregularly timed applications.

Once more we turned to our old standby, neem oil. We considered using it straight or cut with mineral oil to smother the nits, but we chose instead to blend it with a quality liquid soap to emulsify it in warm water, adding a few drops of repellant botanical oils for good measure. On a sunny morning, we grabbed each hen and gently immersed her body in a neem lice and mite drench, thoroughly saturating her skin and plumage before releasing her to dry in the late July breezes. The improvement was almost immediately noticeable, and the lice were gone after two dunks, seven days apart. We plan to use the drench occasionally as a preventative and repellant.

care showed no reduction in red mite populations. Because other options are effective against these insects, we recommend skipping this one.

GARLIC
Studies have shown numerous health benefits of garlic (*Allium sativum*) for chickens, and it's safe to include it in chicken feed. That said, don't overdo it, because if it makes up more than 3 percent of a hen's diet, her eggs will take on a garlicky flavor.

This could be good in an omelet, but it's quite off-putting in flan or crepes. Aged garlic solution, which can be purchased as a dietary supplement, can be applied to wounds on chicks to improve healing, compared to skin lotion or no treatment, so we incorporate it into our homemade antimicrobial wound salve. Other benefits for chickens include a decrease in bacterial count in poop; this reduction in bacteria is likely why studies have shown that garlic supplementation improves food conversion—the amount a chicken needs to eat to gain or maintain her weight. Garlic also kills red mites when applied in two spray applications, but we find garlic too stinky to spray. Although many websites and books recommend garlic for worming chickens, studies have not shown that garlic reduces parasitic worm populations in infected poultry.

LAVENDER

Lavender (*Lavandula angustifolia*, formerly *L. officinalis*) essential oil has been shown to be toxic and repellent to poultry red mites, but its benefits last for only a short time after application, which suggests that repeated applications are needed. We adore the smell, so even knowing that the benefit is transient, we add it to our homemade coop spray—it makes coop cleaning a more pleasant experience. It is also a beautiful garden plant, so it is always featured in our perennial garden. Scatter cuttings in the coop to keep it fresh smelling—your chickens won't mind.

NEEM

Neem is an aromatic oil derived from the neem tree (*Azadirachta indica*). Many studies have demonstrated that a diluted neem solution is highly effective against the red poultry mite, northern fowl mite, tropical fowl mite, several types of feather mites, scaly leg mite, ticks, fleas, bird lice, and flies (including larva). What a workhorse of a botanical! It is the most important component of our approach to the treatment of insects that plague chickens. It has been shown to be safe for spiders, ladybugs, butterflies, bees and other pollinators, and beneficial insects, especially when applied directly to chickens. Studies have also shown neem oil to be a potent antifungal agent. If that were not enough, ½ teaspoon of powdered dry neem leaves (not seed meal) added to every 2 pounds of feed has been shown to be beneficial in boosting chickens' immune systems. Undiluted neem may be used in small quantities directly on poultry skin, but we typically dilute it for our blended spays and drenches. Neem oil spray blends can be easily mixed and sprayed with any size garden sprayer, or a kitchen oil sprayer can be used for pure neem.

OREGANO

This common herb (*Origanum vulgare*) is among the most potent and verifiably effective botanical replacements for pharmaceutical medications. It contains both carvacrol and thymol, two substances with proven antimicrobial and antifungal properties. Remarkably, data suggests that these substances are effective against pathogens such as *E. coli* yet relatively harmless to most other microbes, plants, and animals.

NEEM LICE AND MITE DRENCH

We developed this handy recipe to kill lice, mites, and other external parasites. Use it for outbreaks or as a preventative and repellant. For elimination of lice and mites, reapply one to three times about seven days apart. For a repellant, dunk your hens in the solution once a month or so in spring through fall. It also doubles as a garden insecticide, so don't waste the leftover liquid; use it on your roses or apple trees.

2 gallons warm tap water

½ cup liquid soap (not detergent dish soap) such as unscented Dr. Bronner's

10 tablespoons (5 ounces) neem oil (unfiltered)

5 drops each, clove, lavender, oregano, and rosemary essential oils

Combine the ingredients in large container about 8 inches deep by 12 inches wide. It should hold enough water to submerge a hen's body (not her face!).

Dunk your chickens only on warm days that allow sufficient time for them to dry before bedtime. Catch and subdue each hen by firmly holding her body, thighs, and wings. After she's relaxed a bit, slowly and carefully lower her into the warm solution until she is submerged up to her lower neck. Lift her partway out and repeat the dunk one or two more times. Avoid getting the solution in her eyes—flush the solution from her eyes with fresh water if needed.

Oregano inhibits coccidiosis at a level similar to diclazuril (a synthetic treatment for coccidiosis) when about ¼ teaspoon of oregano essential oil is added to a gallon of feed. This is truly remarkable, especially because we are always looking for suitable botanical replacements for pharmaceutical medications, and coccidiosis is a serious disease.

As with clove oil (except with an even better effect), a solution of 0.25 percent oregano oil (this was the lowest concentration tested) in water reduced salmonella on washed eggs to undetectable levels within 30 seconds. It worked much better than a chorine solution.

PARSLEY

Garden parsley (*Petroselinum sativum*) is often recommended to supplement backyard chickens' diet based on its nutrient content. Avoid feeding your chickens any significant amount of parsley, however. Sharing a little sprinkle of green on some leftover salmon is no problem, but sharing handfuls of parsley that you just deadheaded from your garden is not a good idea. Studies have shown that birds that consume parsley experience increases in photosensitivity, which can result in second- and third-degree sunburns. Spring parsley (*Cymopterus watsonii*), a plant in the same family, grows wild in the Western United States and is considered a threat to livestock because it also causes severe photosensitivity.

PEPPERMINT

Mint (*Mentha ×piperita*) is often recommended by chicken-care sources for its antibacterial, pesticidal, and rat-repellant properties. Although we've often heard about the usefulness of mint as a rat repellent when it's planted around or scattered inside a coop, we cannot find any studies that supports this claim. Indeed, we grow mint and have found its growth extremely difficult to control—and it has had no

CLEAN EGGS

One important thing to know about your hen's unique anatomy is that her gastrointestinal tract intermingles briefly with her reproductive tract (the egg-production chute). This may come as a surprise if you have never given the topic much thought. Yes, both eggs and poop come out of the hen's vent. Gross, right? Wrong! It's ingenious! The lower part of the reproductive tract actually protrudes briefly through the vent as the egg is deposited, thereby protecting it from exposure to chicken waste—as long as you keep your nesting boxes clean. It also coats the egg with the egg bloom, which seals the pores in the egg to prevent bacteria from passing through and contaminating the inside. This amazing and convenient process keeps the eggs surprisingly clean.

Although the egg bloom cleans and protects the egg as it emerges from a hen's vent, eggs can pick up particles of dirt, or worse, from the nesting box (or wherever a hen chooses to lay them). It's usually fine to brush off dirt and use your eggs as they come from the nest. However, if you need to clean an egg to avoid transporting material from a dirty shell into your kitchen, you can wash it in water infused with oregano or clove oil. The water should be slightly warmer than the egg itself: a freshly laid egg that is still warm should be washed in warm water (not hot), whereas a cold egg should be washed in cold water. Keep in mind that washing an egg removes the protective bloom layer from the shell, so it will hasten spoilage. You can partially replace this function and shine up your eggs with a light rub of food-grade mineral oil.

apparent effect on rat populations. Some studies have found mild antibacterial or antifungal properties of peppermint oils, but usually only when it is used in combination with other essential oils. If you like the smell, feel free to add a drop or two to your version of our homemade coop spray. If you are happy to let mint grow freely, it might be rather delightful to smell it as you brush by it to collect eggs. However, if you are like Hannah and have a horror of all things that spread in an uncontrolled fashion, think two or three times before planting it.

PYRETHRUM DAISY

Tanacetum cinerariifolium is the botanical source of pyrethrin, one of the world's most ancient functional insecticides. The plant itself is easy to grow and harvest; once it's dried and ground, however, the homemade pyrethrin compound that kills insects loses effect in twelve to twenty-four hours. Natural pyrethrin can be obtained more conveniently in prepared powders and sprays. Its use is best limited to in-coop application to limit its impact on bees and ladybugs, which never seem to venture in there. If it's used in the run, the chickens' dust bath, or out in the garden, it may kill the good insects you want to keep around. Note that use of synthetic pyrethroids is not consistent with organic chicken keeping. Although they are chemically similar to pyrethrin, these compounds last much longer and potentially pose a more significant threat to beneficial insects. The easiest way to tell the difference is to look for OMRI (Organic Materials Review Institute) labeling, which will tell you immediately whether the product is appropriate for organic chicken keeping.

ROSEMARY

Repellant to mites, but not as effective as thyme, common rosemary (*Rosmarinus officinalis*) smells great and is known to be safe for chickens. Numerous studies have evaluated its effect on poultry as an antioxidant (unimpressive). It is, however, helpful for deodorizing the coop, so we add rosemary essential oil to our homemade coop spray concoction. We always have rosemary in the garden for its use as a culinary herb and for its beauty. If you have extra, it would be safe to scatter rosemary cuttings in the coop to help control odors.

SAGE

One drop of *Salvia officinalis* essential oil mixed in 4 pounds of chicken feed offers numerous health benefits because of its antioxidant properties; though it benefits the entire chicken, it's especially useful in the gut. Tests have shown that it reduces salmonella in the droppings of chickens being supplemented. Sage is another lovely culinary herb that is easy to grow and is ever-present in our garden.

THYME

Thyme (*Thymus vulgaris zygis*) essential oil is toxic and repellent to mites, with a more persistent effect than lavender. When sprayed in a chicken coop, it offers

WHAT *does* ORGANIC MEAN, ANYWAY?

The term "organic" means different things to different people. To a chemist, it refers to anything that is or once was alive (and made of carbon). Ask an organic farmer, and she'll have much more to say. She'll explain how her farm relies on naturally enriched soil, rotations, crop diversity, and natural pest management, and how she adheres to animal welfare standards. They're both correct, of course.

So, what does organic mean to keepers of small flocks? When buying chicken food marked as organic, the label indicates that the feed has been certified by a third party to ensure that its ingredients were grown and processed without the use of synthetic fertilizers, herbicides, and pesticides. When we choose to raise chickens organically at home, we are expressing a commitment (or an intent, at least) to feed our flock organic feeds and use nonsynthetic medicines for prevention and treatment of illness. This doesn't mean, however, that every morsel we provide must bear a USDA organic logo or that OMRI (Organic Materials Review Institute) is going to prohibit us from using an eye ointment on an injured chicken.

For us, raising chickens organically simply means that we do the best we can to employ natural (nonsynthetic) or certified organic care and feeding options whenever we can reasonably do so. Would we use a synthetic pharmaceutical to save a chicken's life if it's the option most likely to work? You bet! Do we buy nonorganic feed when our household budget is tight? Absolutely!

COMFREY: HEALTHY *for* HUMANS BUT NOT *for* HENS

Although comfrey (*Symphytum officinale*) has a traditional use in humans for gastrointestinal disorders, the only published study on use in chickens demonstrates surprisingly high toxicity even at low doses, especially in whole-plant extract form. We would recommend not even planting it in areas where your chickens forage.

antibacterial properties against several types of noxious bacteria. Its effect may be amplified when it's mixed with peppermint oil. It inhibits coccidiosis in laboratory studies, but the herbal oil is not well studied in living chickens. It is a very important component of our homemade coop spray.

NONBOTANICAL NATURAL REMEDIES

Several safe and effective natural options are useful for food, water, and wound care to keep your chickens perky and productive.

APPLE CIDER VINEGAR

Raw apple cider vinegar is often touted as a cure-all for everything from worms to stress, but there is little data to support the various claims. On the flip side, there is also little evidence showing that it can harm chickens, and many observational accounts describe it as a benefit for maintaining appropriate pH in a chicken's crop and gut. A 2012 study showed that a liquid compound of silicic acid and bamboo vinegar was beneficial for the overall health of chickens, leading to improved weight gain. This is the closest thing we could find for data related to vinegar treatment of chickens, but this is largely because it is unstudied. Most resources recommend 1 to 2 tablespoons per gallon of water, but the study showed the greatest benefit at a dosage of about 2½ teaspoons per gallon—so if you use it, this would be the closest we can come to an evidence-based recommendation.

BENEFICIAL MICROBES

These specialized microbial inoculants (such as the proprietary EM blend) have proven to be effective at breaking down ammonia and hydrogen sulfide, the primary foul-smelling and toxic gasses produced by the action of wild decomposing microbes decomposing animal waste. Spray a BM inoculant mixture over the litter and lightly over other coop surfaces with every other cleaning: mix 1 ounce BM in 1 quart of distilled water. In wet Portland winters, when additional moisture is not welcome, we use an even more concentrated solution of about 1 ounce BM per pint (¼ liter) of water and use a little less each time we spray. The EM product that is sold to farmers is essentially identical to the bottled inoculant that you can buy inexpensively, marketed as a soil amendment. EM/BM is also the inoculant used in the bokashi process.

When mixed with feed, EM/BM offers other benefits, including bolstering nutrient availability and intestinal microflora, reducing incidences of diarrhea, and improving egg formation and yolk quality.

BENTONITE CLAY

This clay has been shown to reduce the effect of mycotoxins produced by mold in improperly stored feed. The most common result of mold poisoning is aflatoxicosis, in which the chicken's liver can be severely damaged, often leading to death. Bentonite clay can also help protect against environmental toxins. Chickens in an environment contaminated by mercury absorbed 60 to 100 percent less of the toxin when bentonite was included in their feed. In a pound of chicken feed, adding 1 ½ tablespoons of bentonite can help remove toxins without reducing microbial richness and diversity in the gut.

BIOCHAR

As a food additive, biochar works similarly to bentonite, impacting many pathogenic bacteria while enabling normal gut flora to flourish. As it moves through the chicken's gut, it is ultimately deposited back into the litter as poop, where it will improve the composting properties of the soiled litter (it is also an excellent garden soil amendment). Despite (an in some ways because of) its excellent properties as an antitoxin, biochar is not allowed to be included in chicken feed in the United States because of concerns that it could be used to mask the presence of toxins in the feed.

Unless we were to have a specific concern about toxins in our hens' feed or immediate environment that warranted an increased use of biochar, we prefer to use it in litter or scattered in other areas where our girls will find it themselves. If they eat it, great. If not, they probably don't need it (like free-choice oyster shell or grit).

BOKASHI

For our six hens, we sprinkle a couple of tablespoons of bokashi bran over their daily feed ration three or four days a week to help suppress pathogens and promote the proliferation of beneficial microbes. We also add about a cup of activated bran to their litter each time we add material, and then tamp it down firmly (which the hens soon fluff).

COLLOIDAL SILVER

This suspension of silver is believed to have good antimicrobial effects, especially against staph infections of skin and mucus membranes. It disrupts bacterial biofilms, the protective substances excreted by bacteria that are difficult even for antibiotics to penetrate. Although prolonged excessive exposure to colloidal silver causes argyria, a condition that turns the skin permanently blue, short-term topical use has not been shown to cause argyria or other health issues. Because it may impact the function of healthy flora if taken internally, we recommend restricting its use to topical applications. Most online sources recommending colloidal silver cite the claims from a March 1978 issue of *Science Digest* (a popular magazine), which turn out to be completely unsubstantiated. However, current studies published in peer-reviewed (legit) journals do show benefit.

FOOD-GRADE DIATOMACEOUS EARTH

This natural sedimentary rock comes from the fossilized remains of ancient algae. In gardening and chicken keeping, it is useful as an insecticide that absorbs oils from insects' exoskeletons, causing them to dehydrate and die. (Don't tell any kids who have a soft heart for insects.) Studies have shown it to be effective in controlling internal parasites (worms) when fed to chickens at the rate of 2 teaspoons per pound of feed. Uncalcined (uncooked), food-grade diatomaceous earth has a smaller particle size, with little or none of the crystalline silica that is present in the calcined type used as a filtering aid (mostly in swimming pool filters). The process of calcination has been linked to pneumoconiosis and silicosis in people with chronic exposure. If you buy uncalcined, food-grade diatomaceous earth, it's safe for you and your hens.

HONEY

Manuka honey and honeydew honey have been demonstrated to be helpful in preventing bacterial infection of wounds, speeding wound healing, and even providing antibacterial benefit in sinus washes (at a concentration of about ¼ teaspoon per 5 cups of water). We add it to our antimicrobial wound salve. These honeys differ from typical honey that you find in food sections at the grocery store: manuka honey is produced in New Zealand by bees that pollinate the manuka bush (*Leptospermum scoparium*), and honeydew honey is made from the sticky substance created by insects such as aphids.

PREBIOTICS

These nondigestible food ingredients promote the growth of beneficial microorganisms in the intestines by selectively feeding beneficial bacteria. Prebiotics are often added to high-quality chicken feed, and several (best raw) foods also contain prebiotics, such as jicama, dandelion greens, Jerusalem artichoke, chicory root, sprouted grains, plantain, and peas.

PROBIOTICS

These beneficial bacteria support colonies of healthy intestinal flora. Probiotics are often included in high-quality feeds and are also available as a water additive. The two types that are best for chickens are *Bacillus subtilis* and *Lactobacillus acidophilus*. We make a lovely probiotic bowl for our hens that transforms regular feed ingredients into easier-to-digest rations and also contains a bonus population of beneficial microbes. See page 138 for a recipe.

If you don't want to make your own probiotic, a great commercial product is Big Ole Bird Poultry Supplement, which you add to your hens' water. It makes the water look rather brown, but the birds don't seem to notice or care. Another commercial product, Rooster Booster, has an electrolyte formulation with lactobacillus that is handy if you think your chickens also need electrolytes—if you suspect dehydration, for example.

PYRETHRIN

This natural insecticide is extracted from a species of chrysanthemum flower. (Synthetic pyrethroids are also developed as insecticides and are similar in chemical structure.) It works as a neurotoxin and is highly lethal to insects. As discussed on page 133, we avoid the synthetic types, which do not biodegrade quickly and can present a threat to beneficial insects such as bees. Pyrethrin can be useful inside the coop because mites love to hide in the nooks and crannies during the day. We would not use pyrethrin preventatively (as we use our essential oil spray after each coop cleaning), but if you are certain you have a mite infestation in your coop, you could spray or dust every nook with pyrethrin after a thorough scooping

HENS' PROBIOTIC BOWL

There are many variations to this basic recipe, but we like this simple one that serves about three hens.

2 cups pelleted feed or cracked grains

1 cup warm, dechlorinated water (add ¼ teaspoon humic acid if water contains chloride)

1 tablespoon plain yogurt with live cultures

optional: 1 tablespoon EM/BM probiotic

Mix all ingredients and allow the mixture to stand, covered, at room temperature for 24 hours. Drain off the liquid, and immediately feed it to your hens.

of litter, and then dust with uncalcined, food-grade diatomaceous earth for good measure. We find that sprays are much easier to use for any area above floor level, but powders are excellent for sprinkling into the cracks between boards at the bottom of the coop.

ADMINISTERING DAILY SUPPLEMENTS »

The research we relied on to confirm (or refute) the effectiveness of various botanical and natural supplements is credible in part because researchers used repeatable protocols that could be followed to by other researchers to verify their results. As our own chickens would tell you, if they could, the only constant around our house these days is change. Some mornings it seems we're lucky to remember to feed them at all, let alone add a supplement. To gain the fullest benefit from the practices we recommend, we urge you to be more like the researchers than us in this regard.

Along with maintaining a healthful environment in the brooder, coop, and run that's slightly moist, mildly acidic, and loaded with nonpathogenic microbes (clean and dry also works), we recommend following a daily supplementation regimen for a baseline of immune support.

A daily preventative regimen

+ **Chick water** Replace water daily. In a quart-sized waterer, add ½ teaspoon of aloe powder or ¼ cup aloe juice before filling with clean water. If you think chicks need a little boost of healthful probiotics, add 1 teaspoon of BM/EM liquid microbial inoculant and ½ teaspoon of Big Ole Bird or Rooster Booster with Lactobacillus. Alternatively, you can add probiotics via dry bokashi bran at a rate of 1 teaspoon dry bokashi blended into 1 quart of dry feed. At our store, we provide electrolytes only to chicks or chickens under stress from recent shipping, heat, or illness. Because the large amount of sodium contained in electrolytes may be harmful with long-term use, we recommend that you use them for only a day or two, and use slightly below the recommended dilutions to be safe.

+ **Chick feed** Add 2 tablespoons of bentonite clay and 1 tablespoon of powdered *Artemisia annua* to 1 quart of high-quality chick feed. It's easiest to mix this up in a bowl before transferring it to the feeder; this also ensures even ingredient distribution. If you are not adding probiotics to the water, they can be administered in the food instead. Add either 1 teaspoon of bokashi bran or a single capsule of a human probiotic supplement (ideally containing *Lactobacillus acidophilus* or *Bacillus subtilis*) to the feed and thoroughly mix it in.

+ **Adult water and feed** Continuing the same beneficial regimen will help keep your adult hens in tip-top shape. Just make sure you adjust quantities of supplements proportionate to the larger sized waterer and feeder. Use about 2 tablespoons of *Artemesia annua*, 4 tablespoons of bentonite, and ½ cup of bokashi bran (or two human probiotic capsules) for every 3 pounds of feed. This is easy to mix thoroughly in a larger feeder using your hands in a scooping, turning motion.

MOCK DISEASES

Certain conditions may make you think your chicken is on the brink of death when it is actually just fine and dandy. Familiarizing yourself with these issues will help you know what is normal and what is not.

BROODINESS

At some point, you will count your chickens as they come out of the coop and notice that one is missing. You will open the coop and find the missing lady in a nesting box. You reach to stroke her feathers and suddenly she puffs up—quite literally—with indignation. She may also growl and peck at you. If you bravely persist and search under her, you may find a clutch of eggs.

The best thing to do is boldly shoo her out of the coop and take the eggs. She will be outraged, but when presented with nice fresh food and water, she will usually agree to partake of it before going back to sit in the nesting box. With luck, she will be discouraged by the lack of eggs, but a truly broody hen will often wait around until her friends lay an egg or two and then sit on those.

If you toss a broody chicken out of the coop daily and take eggs away from her once or twice per day, she will usually accept defeat within a week and resume her normal routine. If a hen is persistent, the traditional advice is to put her in a broody box—a drafty, uncomfortable, unsuitable-for-hatching

kind of place that is supposed to convince her sooner rather than later to give it up. Though you'll find many broody box designs online, a dog crate placed in the middle of the chicken run should do the trick—it'll be chilly but will still provide protection from predators.

You may wonder why there is all this urgency about breaking broodiness. The answer is that a broody hen stops laying until she is over it, and in a very persistently broody hen, this can take a month or more. To be clear, you don't actually have to do anything about it. If you can live with one less egg (give or take) per day for the duration, just toss her out of the coop every once in a while so she gets a drink and a nibble of food.

Finally, on a frivolous note, if you've been thinking about expanding your flock, you can use a good broody hen and satisfy her deepest mothering instincts at the same time. Obtain some fertilized eggs, mark them clearly to differentiate them from the other eggs in your coop, and stick them under your fierce mama. The process lasts twenty-one days, and as long as she sticks with it, you should have some baby chicks to show for it. Keep in mind that the chicks cannot fly or hop very far, so the nest setup will initially need to be at ground level. Mama will then usher the

babies around and show them the ways of the flock until they have grown more independent. Typically, she will also protect them from the rest of the flock. Feed mama and babies chick food until it's possible to feed them separately later on. Using a broody hen to help introduce new chicks to the flock can be a great way to expand the flock without enduring much strife or bullying. The only downside is that 50 percent of the eggs will hatch future roosters.

FULL CROP

One anatomical peculiarity of chickens is the crop. Knowing how it should look from the outside can be your first clue to a chicken's health. Chickens, as you know, have beaks rather than teeth, so they need to swallow tiny pebbles, or grit, along with their food. The crop is the temporary holding area before this all gets churned about in the gizzard. When a chicken has a very profitable day hunting insects and other tasty bits in the yard or run, she can end up with an apparent mass protruding just under her breastbone. As long as she looks extremely pleased with herself, rather than ill or listless, she's fine. Expect it to be much diminished or totally gone by the following morning. Think of this as the chicken equivalent of a Thanksgiving gut. She just needs to loosen her belt, so to speak, digest, and sleep it off. If she looks terrible the next morning, and her crop is still full, she may have an impacted crop.

MOLTING

Here's a story we have heard thousands of times:

We got our chickens as baby chicks about a year and a half ago. They started laying late in their first summer, but then the laying slowed down in the winter. We called you then and you advised adding a light to their coop on a timer to give them a few extra hours per night. That worked pretty well, but they really started laying like gangbusters in the spring. It was amazing! But then this fall, all of a sudden they stopped laying again, and the light didn't help, and now they are losing all their feathers! I feel terrible for them! This all started right when it got cold, and they look so miserable and bedraggled out there in the cold. What in the world is going on?

This is one of our favorite stories, because we have nothing but good news by way of explanation. The chickens are simply molting—a totally normal, annual process of feather replacement that all chickens undergo. Generally, they skip molting in their first year of life. The sight of these scruffy-looking hens throws people off, because they think they've kept their chickens really healthy the first year, and then this ugly thing happens. It really is ugly. Those poor chickens.

Opposite, top: Molting isn't pretty, but it's not dangerous.

Opposite, bottom: This chick is in the classic position for sudden chick sleep syndrome.

Egg laying generally comes to a total halt during molting, and hens will not resume laying until molting is done. The molting process generally takes about six weeks, but you can speed the process by increasing the protein in a hen's diet at this time and making sure you are taking all suggested measures to optimize her overall health. The protein can come from your own kitchen (chickens love leftovers). Cottage cheese is universally chicken approved, and even meat is fine—if it is not chicken meat (cannibalism is not okay). Look for protein-rich, non-sugar foods you might have on hand that you are able to share.

Do you dig giving your chickens scratch as a snack? Switch out the carb-heavy scratch during molting for something healthier and even more fabulous (as far as your chickens are concerned), such as dried (or live) mealworms. Yum! If you have been adding probiotics and other chicken supplements less regularly, bring them back. Try adding something new—*Agaricus blazei* mushrooms or black cumin (*Nigella sativa*) in particular.

We always recommend doing a quick chicken exam during molting to make sure there is nothing amiss. Although molting is generally a satisfyingly easy slam-dunk diagnosis, feather loss could be concurrent with an actual illness. Molting should not cause lethargy, lack of appetite, obsessive scratching, or any other signs of a sickly chicken. The patches of bare skin should look otherwise healthy.

When your hens no longer look like they have a frightening skin condition, and they are fully feathered again, by all means bring back that light and get them laying. The upside of the molt is that your chickens will generally look fantastic and glossy when they are done. Any old bedraggled feathers are gone, and any discoloration from poop stains or medical treatments you have used lately will be gone. Think of it as your chicken undergoing a mildly unpleasant beauty treatment for which the price is buying your own eggs for a while.

SUDDEN CHICK SLEEP SYNDROME

We admit it: This is not a real name for an actual condition. It simply describes the scary event when one of your beloved puffballs suddenly closes her eyes, sways a bit, and then keels over on her face and lies still as if dead. If she has been otherwise right as rain, the chances are good that if you give her a gentle prod or just wait a few minutes, she will pop right up to her feet again. One of the characteristics of baby chicks is that, like human children, they are totally unaware of becoming overtired. Also, much like human children, we are unable to convince them otherwise. So they simply fall asleep on their feet. If you are prepared for it, you will likely find it totally adorable.

GASTROINTESTINAL ILLNESSES

A number of maladies can afflict the gastrointestinal system of a chicken. Most commonly they are brief and resolve without treatment, much like the illnesses we humans experience. Several common ailments do require management, however, and we have done the research to determine how you can safely accomplish this while avoiding prescribed antibiotics or the costly care of a veterinarian.

COCCIDIOSIS

As novice chicken keepers, we were startled to discover that the only chick feed available locally at the time was medicated. We later learned that this was a precaution against coccidiosis, a disease caused by a parasitic infection that results in the formation of lesions on the intestinal walls of chicks and hens. This serious disease is often fatal to chicks and can sicken or kill mature hens. In both juveniles and adults, the early signs of coccidiosis can be simple lethargy or fluffed, greasy feathers. Later stages are accompanied by the more well-known symptom, bloody diarrhea, but by the time this is noticed, the infection is well entrenched and often difficult to treat—hence the need for prophylactic medication.

The drug most commonly added to chick feed is amprolium, a coccidiostat that suppresses the growth and reproduction, but doesn't kill, the cause of coccidiosis, a protozoa of the genus *Eimeria*. Our investigation of amprolium was reassuring, revealing it to be widely regarded as both safe and effective. Moreover, the studies we reviewed indicated that, unlike antibiotics, amprolium posed a low risk of drug resistance, particularly when used on the scale of the urban and rural small flock.

Despite lingering concerns about the safety of medicated feed, we fed it to chicks passing through our own store brooders as a precaution against spreading coccidiosis to our customers' flocks. We set a goal of phasing it out, however, but not before a viable natural strategy for prevention could replace it. Knowing that eimeria oocysts (eggs) must be ingested to infect new chicks, we focused on preventing fecal contamination in food and water within our brooders. We worked with a local cage builder to create custom brooders for our store with ½-inch wire floors suspended above pull-out, litter-filled trays. Without any

PROVIDING SUPPORTIVE CARE

When you are caring for an acutely ill or injured chicken, the first (and sometimes only) thing you can do for her is provide supportive care. This means making her comfortable by isolating her from her flockmates in a safe, cozy, and warm (not hot) location, usually indoors. Keep her inside a large, open-topped cardboard box lined with a clean towel, or in a similar enclosure. Provide water, but expect to help with regular dabbing of a mild electrolyte solution on the side of her beak if she's unable to drink. Never squirt fluid into her throat because she may aspirate it. Limit stress and handling when possible, but do use this time to examine her and attempt appropriate interventions.

litter to kick around, the chicks in these special cages experienced no coccidiosis infections, confirming that the best strategy was prevention.

Although the special cages made sense for short-term chick housing at the store, we felt that widespread adoption of wire-floored cages for brooding would not be appropriate. Our main concern was that without litter to dig around in, the chicks' scratching and foraging instincts would be thwarted, putting them at risk for boredom-related behavioral problems such as aggression, which can be nearly as serious as coccidiosis itself.

Unsure of how to proceed, we had little advice to offer customers beyond recommending that they do their best to keep things clean inside their brooders—change litter frequently and elevate waterers to prevent them from becoming a poopy paradise for protozoa. The same advice applied equally well for mature hens, although separation of food and water from the poopiest areas in their environment (under the roosts) in most setups meant deep litter was also a viable litter management alternative for them.

We were delighted to discover that recent research has identified several effective and safe natural remedies capable of replacing amprolium. We have since developed our own preventative regimen of food and water supplementation based on these findings for both chicks and adult chickens (see page 139). Coupled with clean water from nipple waterers and our recommended litter management strategies, these daily additions deliver a knockout punch of potent protection for your poultry!

If, despite taking these measures and carefully managing litter, you notice bloody diarrhea in your brooders, runs, or coops, you need to take treatment to the next level. Make some antimicrobial herbal oil spray (see page 147) and spray it on their feed. If you have no time or patience for creating your own spray, try a commercial product such as Durvet DuraStat with Oregano as a reasonable substitute. It's somewhat less fussy to use because you can mix it directly into the water fount. (We get it. As busy, employed, and over-stretched parents of two, we wouldn't have time for making our own either if we weren't completely chicken-obsessed.)

As your chicks grow into awkward teenagers and finally fully feathered hens, continue adding aloe and probiotics to their water and bentonite to their food, and keep your herbal oil spray on hand (and labeled for its specific use) in case you become suspicious of a microbial infection. Once present in a flock, eimeria is almost impossible to eradicate completely, whether or not you use heavy-duty anticoccidial medication. The illness can crop up periodically, but usually it affects only hens that are stressed or otherwise in ill health. If you have started your chicks out right with the tools their little bodies need to fight illness and build up their resistance, coccidiosis infection is unlikely to occur.

IMPACTED CROP

A chicken's crop can get impacted (backed up) if she has access to long, stringy edibles such as long pieces of grass or pseudo-edibles such as string. A lack of grit could also contribute. Usually, a yard with grass will also contain small stones that she can ingest, so a lack of grit is usually not a concern.

To prevent impacted crop, mow your grass regularly to prevent your flock from eating material that could potentially cause a problem. Also, try not to leave any items lying around that would look enticingly wormlike to a chicken but that are indigestible. We are notoriously lazy about picking up after our yard improvement projects, and it is an absolute miracle that none of our hens has had an impacted crop resulting from eating jute twine.

You should be suspicious of a crop problem if your hen has a full-feeling or very firm-feeling crop when you let her out of the coop in the morning. She should have digested everything she ate the day before by then, and her crop should not be protuberant before she has had anything to eat that day. Also, she is likely to look or act like she's not feeling well.

Impacted crop can be fatal; an afflicted bird generally has no appetite and will stop eating. It sometimes requires surgery, which we do not recommend undertaking yourself. In her book *The Chicken Health Handbook*, Gail Damerow gives instructions for how to perform that procedure if you are confident in your abilities or have some medical training.

There are steps you can undertake to try curing the blockage without surgery, and we advocate trying these first.

Treat an impacted crop

+ Isolate the hen from the rest of the flock, and withhold food until you help her empty her crop. Allow her access to water. Give her 2 teaspoons of olive oil by syringe. Do not jet it down her throat, because it might go down her trachea instead and cause respiratory problems. Instead, let it dribble in; she will usually swallow

ANTIMICROBIAL HERBAL OIL SPRAY

If you suspect any significant illness in your flock, you can make an antimicrobial formulation and add it to your chickens' feed. This spray uses an effective organic antibiotic alternative using essential oils in a carrier of olive oil. When added to feed, it provides immune system support for hens by targeting protozoal and bacterial colonies. It's especially useful when combined with regular probiotic supplementation because the good bacteria they contain seem to be unaffected by the active ingredients in the oils. Take that, germs!

1 cup olive oil

¼ teaspoon oregano oil

2 drops lavender essential oil

2 drops thyme essential oil

1 drop sage essential oil

Mix all ingredients in a quality spray bottle that is capable of spraying thick oil. Add 1 quart of feed to a mixing bowl and directly spritz the feed once or twice (one large spritz or two small ones) with the bottled mixture and then mix it in. In large feeders, for every 1½ pounds of feed, spritz once or twice in the feeder itself and mix it in with your hands. Continue treating the feed until signs of illness have been gone for at least a week.

the drops reflexively. After this, gently massage her crop in a top-down pattern for about ten minutes. With luck, by the end of this process, the crop will feel considerably less distended. Check on her every morning for the next two days, and repeat the process if necessary until she wakes up with an empty crop. At that point, you can slowly reintroduce food, starting with easily digestible chick mash.

➜ If her crop issue is not improving after two days of this regimen (or earlier if she is becoming lethargic), contact your vet. Home surgery might be an option if you simply cannot afford or justify the expense of a vet visit for a chicken and it would otherwise mean losing her. Remember, though, that a surgery affecting the gastrointestinal tract of any creature carries high risk of sepsis—a terrible way to die.

PASTED VENT

A chick can develop a condition called pasty butt, sticky droppings, or, most politely, pasted vent, in which poop adheres to her bottom, causing a plug to form. This literally prevents her from pooping and can quickly lead to her death if not treated. This does not happen in adult hens and usually occurs only in the chicken's first two or three weeks of life.

Pasted vent is thought to be caused by stress and/or dehydration. In the United States, most chicks are shipped in boxes from big hatcheries when they are a day old, and they arrive at feed stores having never had anything to eat or drink. We suggest you try to prevent pasted vent by feeding your new chicks water supplemented with probiotics and electrolytes from the moment they arrive. This is doubly important if they have arrived in the mail and have visible signs of stress. Hatcheries sometimes offer to package chicks they ship to you with a snack consisting of seaweed gel, sugar, and electrolytes to reduce shipping stress, and we recommend opting for it. It's never too early to include aloe juice or powder for coccidian control. After a few days, discontinue the electrolytes because the sodium it contains is no longer helpful.

Some studies have shown a reduction in sticky droppings after giving chicks digestive enzymes. However, the enzymes they get depend on what is in the feed you are using, and not every backyard chicken keeper can analyze the feed and then coordinate enzymes.

We think you'll find it more cost-effective and beneficial for the chicks to check them for problems until they are out of the danger window. If you purchase or hatch chicks, keep an eye on them by picking up each chick once per day, turning her over, and examining her vent. Keep in mind that a very young chick will also have a visible healing umbilicus near her vent; don't make the mistake of thinking this is a pasty vent and picking at it.

Alternatively, if you get chicks from a feed store, look for slightly older girls. Chicks that are a few weeks old are not as darling as the day-old puffballs, but they have survived past the initial danger zone of stress and adaptation after shipping, and this reduces the possibility of pasted vent.

Unplug a vent

+ Holding the chick gently but firmly, apply a warm, wet washcloth or paper towel to her bottom and hold it against her for five minutes. Usually, she will happily settle into the warmth, often even closing her eyes and falling asleep. After this, check to see if the plug can be gently wiped away.

+ If it isn't budging, don't pick at it. Apply the warm cloth for an additional five minutes, and then try again. If it still won't come off, look closely to determine if there is a bit of fluff clinging tenaciously to the plug. In this case, very carefully use a pair of fingernail or thread scissors to cut the bit of fluff away, which is infinitely more comfortable to the chick than pulling it out.

Left: Don't confuse her healing umbilicus (that little pink spot on her belly) with her healthy pink vent (just under her little tail).

Right: This chick is perky enough, but the clump of poop stuck to her bottom is a pasted vent, and it can be fatal if not addressed.

✦ If all else fails, hold her little bottom under a gentle stream of warm water until the plug is visibly melting away. This is not our preferred approach, however, because it is a bit more startling to the chick, and you run the risk of inconsistent temperature and water pressure, both of which could be harmful if they were to change abruptly.

✦ Once the vent is cleared, if you caught it early enough, your chick should be in the clear for the day. However, keep a sharp eye on her, because a pasted-vent chick is prone to repeat incidents. Treating pasted vent once a day will be sufficient to keep the chick healthy.

PENDULOUS CROP

Some of our customers who are lucky and wise enough to have kept their girls alive for a long time have very aged chickens who struggle with pendulous crop. In this condition, the crop seems to have little support and hangs limply under her breastbone. It is very rare in a young chicken, but in some cases individuals have noted a genetic aspect, in that chicks from a hen with pendulous crop are more likely to have it themselves. Sometimes, this causes no symptoms other than a drooping crop, but it can make the chicken prone to sour crop or impacted crop, likely because of the abnormal movement of crop contents through the gastrointestinal system. In severe cases, the pendulous crop can restrict a chicken's movement as it drags on the ground, and the skin can be denuded of feathers in this way.

We also suspect a bit of a chicken-or-the-egg dilemma here (forgive us). A prolonged or recurrent issue with sour or impacted crop would certainly increase the likelihood of a pendulous crop developing because of the increased weight of the crop over time. Conversely, pendulous crop makes sour or impacted crop

more likely. The condition is associated with older chickens, and it was rarely seen in flocks until chickens became valued as pets and started living longer than traditionally farmed birds. Chickens on farms rarely live for more than a few years, as they are culled as soon as their egg productivity begins to decline. When they live pampered lives in lush backyards, hens can live to ripe old ages, have access to more problematic long grasses, and develop syndromes associated with age that were rarely problematic in the past.

A chicken with a pendulous crop needs to have it supported by external means. In short, you need a chicken bra. Bless human ingenuity, because some folks out there who are handy with a sewing machine have made these and are selling them online. Or, if you know how to work a needle and thread yourself, then by all means, make your own chicken bra, and make it as fancy as you like. The design involves a padded rectangle of fabric suspended between two long elastic straps that tie behind the bird's back. Some have had luck using an athletic sock slipped over the hen's head with holes cut out for their wings. Just remember that nothing should droop over the vent, which would be messy beyond words.

SOUR CROP

Sour crop can be a bit difficult to differentiate from impacted crop, because its main symptom is also a full crop in the morning. The difference is that the crop usually looks full but feels soft, and your chicken may have terrible breath. If you pick her up too briskly, she may also expel foul-smelling fluid from her beak. Sour crop, like impacted crop, can be life-threatening, and chronic sour crop will cause thickening of the crop wall, which in turn leads to a predisposition for impacted crop.

Above left: Matilda's pendulous crop is evident. What distinguishes it from sour crop or impacted crop is its persistence.

Above right: Matilda elegantly models her crop bra, which helps alleviate symptoms from her pendulous crop.

Sour crop is actually a yeast infection of the crop, usually caused by *Candida albicans*—the same yeast strain that causes issues with humans. Backyard chickens usually get yeast infections from drinking contaminated water or from having access to a large a quantity of inappropriate foodstuffs, such as moldering fruit. Commercially raised chickens can have yeast problems that result from prolonged use of antibiotics.

It can be prevented by supplementing with probiotics and prebiotics as part of a daily regimen, regularly refreshing your flock's water, and keeping the chickens' areas cleared of fallen fruit. We love to give our chickens leftovers that seem too good for the compost heap, but we always make sure the food is not at all moldy.

Treatment of sour crop can be problematic. The traditional cure is to turn your chicken upside down, and then empty her crop of the foul stuff by briefly massaging it. This usually takes only moments, but it carries a serious risk of causing pneumonia because your chicken may aspirate (inhale) some of her foul stomach contents.

We prefer an approach of increasing probiotic support while treating the fungal infection with a gentian violet and olive oil purge. Yes, the gentian violet says it's for external use only. Yes, it contains alcohol. However, because it is a medication recommended by mainstream pediatricians for treatment of thrush in a newborn human's mouth, we are not concerned about using a scant amount to save a chicken's life.

Gentian violet and olive oil purge for sour crop

+ Add 1 drop of gentian violet to 2 teaspoons of olive oil, and gently administer it with a syringe, letting it trickle down the hen's throat rather than squirting it in forcefully. Don't massage her crop, because this may cause her to aspirate. Repeat this process daily for 1 week. You may or may not notice the side effect of blue diarrhea resulting from the gentian violet. Also, gentian violet stains like the dickens, so don't attempt this procedure right before going to work in your favorite office attire.

+ Assess her progress by her willingness to eat and drink and the appearance of a normal crop. This is not a contagious condition, so there is no need to isolate the unfortunate hen from her flockmates.

+ As she begins to eat and drink more, double the amount of probiotics in the flock's water or feed for an additional week after you stop the gentian violet treatment. This can also benefit the rest of the flock, who may have been eating the same moldering fruit.

+ Mix in a cup of neem meal for every 40 pound bag of chicken feed for the next couple of months, which should provide some long-term antifungal protection.

VENT GLEET

Several young hens at the store appeared to be suffering from persistent, yeasty-smelling diarrhea that would not resolve with our routine preventative regimen. We later learned from a vet that we were seeing vent gleet, a type of fungal (yeast) infection of the reproductive and digestive system.

Treating vent gleet

+ Begin with an Epsom salt bath, followed by an application of gentian violet externally on the irritated vent.

+ Next, go after the fungus itself by adding probiotics, neem meal, and antimicrobial herbal oil spray to their feed (or add DuraStat to drinking water). Use 1 tablespoon of neem meal and a pump of the spray per 8 ounces of slightly moistened mash feed (or add DurVet to the water according to package instructions).

+ Replace the feed daily to prevent spoilage and continue the treatment for seven to ten days, including at least three days after symptoms have improved, to thwart relapse. Continue the use of probiotics indefinitely.

EGG-RELATED CONDITIONS

One of our very favorite qualities of our chickens is that they lay the most delicious and beautiful eggs we have ever seen. This alone gives them an advantage over our other pets, whose byproducts are not nearly as charming. However, the anatomy that enables the chicken to lay her eggs—the oviduct—is not without failings. When a hen's oviduct goes wrong, you will need to know how to recognize the problem and what to do to get her through it.

EGG BINDING

Egg binding occurs when a hen is unable to pass an egg through her vent. The cause for egg binding is usually dietary—either the chicken is eating an unbalanced diet, is not getting adequate nutrients (especially calcium), is dehydrated, or is overweight from too many carb-heavy treats. If egg binding occurs, reevaluate what you are feeding your chickens, make sure they have plenty of calcium in their feed and/or access to sources of calcium, and consider switching from a grain-based scratch to something protein-rich for treats. After having to deal with it once, we promise you will not want to encounter egg binding again.

Egg binding can be fatal, and it needs to be addressed quickly but cautiously. Consider it an emergency situation. Your first clue to diagnosis is often noticing that your chicken is acting strangely, appearing uncomfortable, and not participating in normal chicken activities. She will usually refuse to eat or drink and may have a puffed-up appearance, which is a normal chicken response to pain. She may also exhibit unusual behavior, such as moving repeatedly in and out of the nesting boxes or squatting and straining as if constipated.

At first you may be perplexed, because she will have no obvious external signs of illness. However, as you perform your beak-to-toe exam, you will often be able to feel a mass on her lower abdomen, usually just inside her vent. Examine the area with caution, because your goal above all is to *avoid breaking the egg*. Broken shards of shell can injure the delicate tissue in the lower oviduct.

If you do not feel an abdominal mass, but you can find no other obvious source of her discomfort, it would be wise to perform an internal examination to feel for the possibility of an egg a little further up the reproductive tract. When you do the exam, wear disposable gloves and use lubricant. This will make the exam easier, and it will also potentially lubricate the cloaca and vent to ease passage of the egg. You can use vegetable oil, petroleum jelly, or KY jelly. Any non-irritating lubricant you have on hand will do—just use plenty of it.

Help a hen pass a stuck egg

✦ Insert one finger into the vent and aim upward, along the angle of the chicken's spine, rather than forward along the abdomen. (Consult the anatomical diagram and the reason for this will be clear to you.) You need to access the reproductive tract rather than the gastrointestinal tract, both of which end at the vent. If you can feel an egg within about an inch of the vent, you have confirmed your diagnosis.

✦ While your lubricated finger is in there, gently rub the lubricant over the surface of the inner oviduct and whatever part of the egg you can reach. If she appears to have a lot of swelling of the cloaca from straining, consider applying some Preparation H to reduce the inflammation, which can contribute to obstruction. Sometimes this stimulation plus the lubrication is enough to cause her to lay the egg. If so, celebrate! You are getting off easy.

✦ A relaxing Epsom salt bath often helps a chicken lay a bound egg, and this should be your next step. There is no risk to doing this, and it works about half the time. Afterward, give her some water supplemented with electrolytes and encourage her to drink, but don't offer food.

✦ Then place her in a dim (but not totally dark), quiet area for about an hour. Check to see if she has passed the egg. You can repeat this entire procedure of applying lubricant and an Epsom salt bath two more times over the next day or two, but if it is not effective, you are looking at a more serious situation. This is the point at which you should consider calling a veterinarian. Your only recourse at this point, breaking the egg, can be dangerous, so you have to weigh whether you are willing to accept an increased risk that your chicken could die by your trying to treat her at home.

+ Some resources suggest giving a calcium supplementation orally to help an egg-bound chicken, but we believe that is likely too little, too late. A calcium injection from a vet may be effective in stimulating her muscles to expel an egg, but a bound egg can be fatal within just days, so oral calcium supplementation has little chance of helping—plus an egg-bound chicken will resist eating anyway.

+ As a last resort, break the egg inside the hen. Call your vet before taking this step. This last-resort treatment option carries a very high risk of killing the hen, because of the possibility of internal bleeding or infection, especially if the broken eggshell cuts her. If the egg is fairly far up the reproductive tract, a broken shell will be mixed with sticky egg contents, and the oviduct will collapse around the shell, making it very difficult to remove entirely without incurring serious injury. It just seems like a terrible end for a chicken.

+ If you are inclined to try this despite our warnings, you will need to have an appropriate hypodermic needle on hand. If you are able to see the egg through the vent, you can use the hypodermic needle—the lower number the gauge the better—to pierce and draw out the contents of the egg. Then, with luck, you can collapse the egg gently so it's still held together by its internal lining membrane. If the membrane holds, it increases the chances of your getting the shell fragments out without injuring the hen's tender internal tissues.

PROLAPSED VENT

This is a condition in which the cloaca, which is the terminal end of the oviduct, prolapses out through the vent. The cloaca is designed to do this briefly for clean delivery of the egg, but the problem occurs when it does not return to its intended position. This is most common in young chickens that begin to lay early. This precociousness might seem a good thing to a chicken keeper eagerly awaiting fresh eggs, but young chickens also often lay abnormally large or even double-yolked eggs, which may be too much for their immature reproductive systems to handle well. Prolapsed vent can follow the laying of an overly large egg in a hen of any age, or it can occur spontaneously, but it is most often seen in very young layers or in hens of advanced age.

The key to successful treatment is to identify the problem as soon as possible. If you notice a pink bulge on your chicken's bottom or a clump of poop stuck there, reminiscent of plugged vent, catch your chicken and check her out right away. A prolapsed vent can be a target for pecking by flockmates, and this can cause serious injury or infection, which of course will dramatically increase the severity of the condition. (The other chickens are not being malicious. A chicken's way of exploring something new is to peck at it experimentally, and like many other birds, they naturally have a hard time resisting anything that is bright red or pink in color.)

If you see no other injury to the vent, your goal is to isolate the hen immediately, get the vent back in place as soon as possible, and discourage laying for the next few days by keeping her in a dark place.

Replace a prolapsed vent

+ You will be glad to have disposable gloves on hand for this procedure. Remove any clinging poop as gently as you can. If there seems to be poop stuck just inside her prolapsed vent, gently press on the sides of the vent to help expel it. Then cautiously feel her abdomen for any evidence of an egg stuck just inside the vent. If you feel an egg, read about egg binding, and return to this section when that life-threatening situation is resolved.

+ If no egg is detected, apply a thin layer of Preparation H to the entire prolapsed area, and then gently attempt to push it back into its correct position. Think of aiming up along the chicken's backbone. Often this is surprisingly easy, and the hen seems to experience immediate relief. Do not push too hard, however: the last thing you need is a puncture wound.

+ Whether your attempt is successful or not, the next step is to give the hen an Epsom salt bath. After the bath, apply Preparation H again to any part of the area that is still protruding. If the prolapse is still present, or you were unable to replace it, try one more time to ease it back in. Then finish her spa treatment with a blow-dry. The ointment will help protect the prolapsed tissue from drying out, but avoid pointing the dryer directly at the vent, and keep the heat on low.

+ Enclose the chicken in a small area, such as a kitty carrier or a cardboard box with plenty of breathing holes. Keep her in a dark area, such as a bathroom with the lights off, for two or three days. Offer her water a few times per day, but limit her feed intake to about ¼ cup per day.

+ This isolating treatment might sound cruel after everything we've told you about chickens being social creatures. However, a chicken with a prolapse is not going to feel well. Her best hope for a permanent cure is to stop laying for a few days, and the best way to slow or stop a chicken from laying is to restrict food and light. Also, a sick chicken does not seem to mind isolation, and a chicken with prolapse is in danger around her pecking peers.

+ If you don't find the prolapse until after your hen has already been injured by pecking, follow our recipe to create antimicrobial wound salve (see page 168) and apply it to the broken skin. Then apply the Preparation H to the whole prolapsed area. Alternatively, apply Vetericyn, manuka honey, iodine, or colloidal silver—whichever you have on hand. These will all prevent infection until you can get the ingredients together to make the wound salve.

Prolapsed vent is a potentially life-threatening issue, so if this chicken is your absolute favorite pet and you don't feel confident in your ability to treat her yourself, consult a veterinarian immediately.

THE MYSTERY *of the* MISSING EGGS by Robert Litt

Holding the lid to the nesting box halfway open with my right hand, I reached into the darkness with my left and felt around: wood shavings. I opened the lid wide and the sunlight confirmed what my hands had reported: wood shavings. No eggs. I bent my knees, cocked my head, and peered through the nest box into the dark coop beyond. For a moment I could see only darkness. My eyes painfully adjusted, and I could make out two things: wood shavings and cobwebs. No eggs.

It was midday in June, and the nest box should have contained one light brown, pointy egg; one and possibly a second green-tinted egg; one pale blue egg; and at least one egg so brown you could hide it in a box of chocolates. Closing the lid, I walked around the coop and approached the hens' enclosure, expecting it to be as empty as the nest box. A startled hen turned her head to the side and looked at me with a single, wide-open eye. She froze.

"Where's everyone else, Sweet Baby Chicken? Where are all the eggs?"

"Braawwwk...aawwk," came her reply, which, if I'm not mistaken, loosely translates as "I'd really like to eat your toe." Hearing the squawk, Creamy emerged from beneath the coop, coated beak-to-tail in what appeared to be powdered cocoa. Out popped another hen, and then another, and pretty soon the girls were all there, puzzled and dusty from their bath.

I lifted my hat and scratched my head. I watched as the girls shook off their dust and started casually pecking on the ground, clearly disappointed my toes were not on offer. I began the mental checklist. It had been a really gray, soggy spring, sure—but it hadn't been too chilly, and it certainly wasn't too hot yet. These hens

were all pretty young, but not *that* young. They'd been eating—well enough, anyway.

I looked at their feathers and pronounced them to be perfect: glossy and fresh, with a just a hint of iridescence. I took stock of their rich crimson combs for a long moment before moving my gaze down to confirm the rich pigmentation of their shanks. Check, check, and check. Everything was right as rain; they just weren't laying.

THE CLUES

Judging by the volume of questions we get at the store, I am not alone in this experience. Because egg production is a function of many factors, we usually ask lots of questions before making recommendations. In the final analysis, there is seldom a single cause of egg decline, but after a brief interview with the customer, we are often able to identify a few important limiting factors and make suggestions that can improve yields and quality. So I asked myself the same questions I usually ask customers.

➤ How old are the hens?

➤ Are they molting?

➤ What breed(s) do you have?

➤ How are they eating? Has there been a change in the type of feed used or in how much feed the hens consume daily?

+ What are their sources of light and warmth?

+ What are their housing conditions?

+ What is shell quality like?

+ What is interior egg quality like?

+ Are there any signs of parasites?

+ Are there any signs of disease? Are the birds overcrowded?

How old are the hens?

After hatching, a young hen enters a period of rapid physical development. For some customers it's not rapid enough, as we hear when they come in to buy yet another bag of feed for their "slacking" young flock that has yet to produce anything. Finally, by five to six months of age, hens reach sexual maturity and laying may begin. This is by no means certain.

If it's not a problem of hens that are too young, it's often an issue of them being too old. Although backyard chickens may live four to eight years, they typically lay abundantly for less than half that time. After their second laying season, egg production usually becomes sporadic or may even halt. The backyard chicken keeper then has a difficult choice: continue feeding and caring for aging, unproductive hens, or thin the flock. For some of us, it's an easy decision: we treat them as pets, and there's no question we'll be feeding them long past their heydays. For others, hens that no longer lay efficiently must be removed.

Recommendations Regularly introduce new hens to the flock to maintain a diverse spread of ages and laying abilities. For backyard flocks, we recommend beginning with three to five hens and adding two to four more every other year. Given typical life expectancies, your flock should stabilize at six to twelve hens, with two or more always in the prime of their egg-laying careers. This is a larger flock than most people initially have in mind, but if you want to have a relatively steady supply of eggs through the years, it's practically unavoidable. This is one of the reasons we recommend that you invest in (or at least plan for) a large coop and run.

Are they molting?

By their second autumn, chickens will experience their first molt, an annual renewal of plumage during which laying will cease temporarily. For some hens, feathers will be only slightly ragged, but most will look terrible, with patches of bare skin and new feathers, like porcupine quills, emerging through the tattered remains of old ones. The process usually takes about six weeks to resolve, but by then, light levels are low in most parts of the country and laying will not resume unless you provide artificial light.

Recommendations Provide extra protein in the form of grubs, meat, dry cheese, or other suitable sources to help speed your hens through the molting process.

What breed(s) do you have?

Wild jungle fowl lay five to twenty eggs during each of their two reproductive seasons a year, even in the warm, semitropical climates where they evolved. After many thousands of generations of selective breeding, and perhaps a mutation or two, even the most casual of their domestic cousins produce several times this much annually.

Dozens (if not hundreds) of common breeds are kept in small flocks throughout the world. Until fairly recently, most were bred as dual-purpose or utility breeds that would lay reliably and produce some meat, and most could live well in a wide range of environmental conditions. A good example of this sort of chicken is the Plymouth Rock—hens are steady producers of eggs, averaging about 140 a year over their three best years, and roosters slowly put on a good deal of weight over the course of a spring and summer. The Plymouth Rock is also capable of foraging some of its own food, a trait that's highly valued by small flock keepers. This utility comes at the expense of efficiency, however. Modern hybrids, such as the California White, are highly specialized layers that produce marvelously when conditions are fine-tuned to their needs, regularly laying more than 300 snow-white eggs per year.

Commercial laying hybrids such as the California White will also lay plenty of eggs in a home flock, but for the highest levels of production, these and many other white-egg–laying breeds require a light- and climate-controlled environment and a meticulous feeding program to reach their production potential. In a backyard setting, California Whites are known for being prima donnas, high-strung, and flighty birds that are harder to tame and more likely to escape than heavier, brown-egg-laying breeds.

Recommendations If eggs are the only reason you are keeping chickens, it may be a good idea to choose a homogenous group from a highly productive, modern laying breed and be prepared to maintain a controlled environment by heating their coop in cold weather and providing light. If you are keen to harvest lots of eggs but prefer a hardier breed, consider sexlinks, Ameraucanas, or top-laying heritage breeds such as a Rhode Island Red or Plymouth Rock. Feel free to mix breeds in the flock, as they have similar needs. If eggs are only a bonus and you seek hens primarily for pets or for appearance, there are many breed possibilities and combinations.

How are they eating? Has there been a change in the type of feed used or in how much feed the hens consume daily?

A hen should be eating 4 to 6 ounces of a balanced feed ration per day: any deviation in her eating habits, up or down, should be a warning sign. Increases can be caused by cold weather, parasites (internal or external), onset of laying, and several other factors. Decreases are usually a result of overfeeding treats (you know who you are!), but they may be caused by other stressors in the form of weather, social forces, or disease.

Changes in feed formulation, freshness, and even variations between batches of the same feed all can make a profound difference in laying rates and egg quality. How you feed is also very important, because feed that remains in feeders for too long will become stale and possibly moldy. Similarly, providing fresh water is also important to avoid toxins and disease.

Old feed can also lose some nutritional quality because of degradation of vitamins. Hens rely on certain vitamins, particularly D, for proper egg development, and they will suffer if their food lacks it and they are unable to produce it themselves.

Recommendations Cut back on low-protein treats, continue using the same brand of feed, and buy feed from suppliers that frequently restock from reputable mills. To minimize spoilage, store feed and treats carefully, offer smaller meals more frequently, and/or invest in (or make) a treadle feeder that keeps your hens' food covered and fresher longer.

Change water a minimum of once a week and more often in warm weather. Consider adding boosters—supplements of minerals, vitamins, botanicals, and probiotics—especially during stressful times. Replace your traditional fount waterer with a nipple waterer.

What are their sources of light and warmth?

Light and warmth are key environmental factors that influence egg laying in domestic chickens. Hens' eyes are particularly sensitive to variations in light intensity, and they possess an extra-retinal photoreceptor that is sensitive to different wavelengths of light, which may help them to anticipate seasonal changes.

In most of the United States—in the northern states in particular—the hours of daylight and intensity of light levels drop noticeably in winter, and shading from trees or buildings, lack of windows in coops, and other factors can make this problem more pronounced.

One reason for seasonal decline in egg production is lack of vitamin D. Hens can manufacture this essential vitamin in unfeathered skin cells on their comb and legs if they are exposed to sufficient light, but they rely on their feed to provide it otherwise. Feed usually provides only a minimal amount of vitamin D, however, which may not be sufficient for hens under other stresses or those particularly sensitive to the vitamin's effects.

Along with the lower light levels of winter come lower temperatures. Though most domestic hens of well-adapted breeds should have little trouble enduring temperatures well below freezing, by about 55°F (13°C), the cool weather will begin to affect their metabolism and laying. Conversely, as temperatures rise above 80°F (27°C), laying will also begin to decline. These effects can be so profound that hens may stop laying altogether for days after a cold snap or a hot spell.

Recommendations Begin by positioning the coop, run, and foraging areas where your hens will receive plenty of light, ideally from dawn to midafternoon, with late-afternoon shade during the hottest hours. Your coop should have windows facing east to capture the all-important morning light, which seems particularly beneficial for laying. Don't overdo it: if you live in a hot climate or on hot days, be sure to provide plenty of water and allow your girls access to shady, moist soil if possible.

It's also perfectly acceptable to add supplemental lighting to your coop in cold weather. While Hannah was pregnant, she craved eggs, so we kept the hens laying by using a chick heat lamp in the coop from after their molt was completed in November until the return of spring sunlight. In retrospect, the hens actually seemed to be healthier overall that winter than they did in subsequent years without supplemental light, though other factors were in play. If you want to try using a lamp, safely mount either an infrared chick lamp (which also provides heat) or a regular bulb near the perches and set it on a timer from 4 a.m. to 8 a.m. to get photons on those sleepy peepers. Hens that forage and free range will get more light exposure and may not need as much, if any, supplementation.

Providing a little heat during winter (such as from a brooder bulb) can also improve laying and the overall condition of your flock. A clever device called a Thermo Cube automatically powers any device plugged into it on and off within a range of temperatures. For instance, we've used one connected to a heat lamp in our coop that turns on at 35°F (2°C) and off at 45°F (7°C), efficiently keeping our hens cozy when they need it most. It's available in other temperature ranges, including warm-weather models that can be used creatively for cooling.

What are their housing conditions?

The comfort zone for an adult chicken is between 45°F (7°C) and 80°F (27°C), with humidity below 75 percent. In these conditions, a hen can easily regulate her normal body temperature of 105°F (40°C) to 107°F (42°C) by releasing heat through her skin and by conserving warmth with insulating feathers and fat and by increasing her metabolism. Older, overweight, ill, and younger chickens will be the first to show signs of stress beyond this temperature range, but healthy adults are hardier and can withstand more adversity.

Recommendations Though some breeds have been developed to withstand extreme local conditions that regularly exceed the comfortable range, it's wise to take some precautions during hot or cold weather events, particularly cold that's accompanied by high winds or heat coupled with high humidity. Hot weather seems to stress urban hens more than cold. We suspect that's because so many pet hens are overweight, making them more susceptible to heat-related conditions but insulated from the cold.

A few heat precautions

+ Offer cool water and fresh food in the mornings to encourage appetite. A timed light may be helpful to stir them early.
+ Provide access to moist soil and dusting areas in the shade to aid heat exchange with cooler layers of soil below the surface. Mist sprays and cool pools of shallow water can also help.
+ Avoid feeding hens high-carb treats and switch to something fatty as a source of energy that's not as warming.
+ Add electrolytes to drinking water to help hens stay hydrated, maintain vitamin and mineral balances that aid bodily responses to stress, and help remedy loose droppings that accompany excessive heat.
+ Watch for alkalosis, a serious condition that can result from too much panting to cool off. The blood becomes too alkaline, and the remedy is a counterintuitive alkaline solution of ¼ cup of baking soda in the water to trigger a corrective response by the hens' kidneys. This will also help eggshell development and egg production. Baking soda should not, however, be a long-term supplement because it can cause metabolic alkalosis.
+ If during a heat spell a hen seems very sluggish, has foam around her beak, or is otherwise leaking

fluids, immediately bring her into a cooler area such as your basement or air-conditioned home and dab the baking soda–and–water solution onto her beak. Do not pour or squirt it in her throat or she may aspirate the solution. A cool bath with or without Epsom salts may also help bring down her core temperature.

A few cold precautions

+ Use an electric poultry fount or place a regular metal fount on its matching heating base to keep drinking water from freezing. Or bring the fount indoors each evening and, come morning, return it to the coop, where it will remain liquid for several hours before refreezing—ample time for hens to hydrate themselves. Cautiously provide water inside the coop if deep snows prevent your hens from exiting.
+ Add an infrared heating panel to the coop to provide an efficient source of warmth for your hens. Heated perches also provide warmth directly to exposed toes and hens' bellies. The alternatives include heat mats and lamps. Never use electric space heaters intended for indoor use in a coop!
+ Cold weather is one time it's okay to give hens a few extra corn scratch treats: there's less concern about diluting protein because laying is lighter and the sugars will help stoke their metabolic furnaces through the night.
+ Larger flocks of hens (four or more) will retain and share warmth better than smaller groups. Consider adding a bunny to help keep it cozy!
+ Provide quality shelter; this is essential—four walls, a solid floor, and a roof at a minimum. Insulation helps, but heavily insulated coops can also trap moisture, so always provide some ventilation above the hens. Sleeping hens are most at risk during cold weather.

What is shell quality like?

We are motivated to keep laying hens not in hopes of the abundance of eggs, but for egg quality. We want to see clean, thick shells without wrinkles or bumps and in the appropriate color and size for our breeds.

Hens that do not ingest enough protein may lay small eggs or even stop laying altogether, but don't be overzealous about feeding them too much protein, because it can result in excessively large eggs that can lead to egg binding and death. It's better to cut back on snacks than to overfeed protein.

Various egg defects have common causes and solutions. It is now widely accepted by poultry professionals, for example, that giving hens only the minimum levels of vitamins adequate to prevent an actual deficiency is not enough to support their optimal health. Not only are hens capable of using higher levels of vitamins and minerals, but a variety of factors from stress resulting from inefficient feeding may increase their baseline needs. In other words, their optimal intake is dynamic, and it's our job as their keepers to respond to their changing needs.

Eggshells may display a variety of defects, from hairline cracks to shells so thin they look like plastic bags. A young hen's first few eggs may have abnormal shells (or no shells at all) and may be small, but this is considered normal.

Problems leading to shell defects

Calcium A deficiency of calcium is the most common cause of thin and cracked shells. Eggshells are comprised of 95–97 percent calcium carbonate crystals, and it's vital that your hens receive about 4 grams (about ⅒ ounce) of calcium daily. Providing each chicken 4 to 6 ounces of pelleted layer feed usually provides enough calcium, but if forage, kitchen scraps, or treats make up a portion of your flock's diet, you can sprinkle a little calcium over their feed. Or give your hens oyster shell, a source of calcium that is commonly available at feed stores. As usual, don't overdo it: excess calcium in the diet, especially in hot weather, can lead to alkalosis, a cause for far more concern than a few thin eggshells. Excesses can also interfere with phosphorus balance and absorption, reduce potassium uptake, and lower blood oxygen levels. Feeding adult levels of calcium to immature hens can interfere with bone formation, damage kidneys, and (in extreme cases) cause death.

Magnesium Excessive magnesium can interfere with calcium uptake, so avoid supplementing feed with dolomite lime, which contains lots of this mineral.

Salt Excessive salt can cause cracking in eggs. Astonishingly, studies have shown that salt can cause irreversible impairment of normal shell formation, even at relatively low levels. For this reason, we recommend that you avoid combining baking soda and electrolytes. This also means no feeding your hens pretzels or other salty snacks!

Vitamin D3 Even in the presence of calcium, a hen requires vitamin D3 to be able to absorb calcium into her intestine. Most feeds contain this vitamin, but high storage temperatures, long periods in storage, and poor mixing may reduce its potency. Hens can make their own vitamin D3 from exposure to sunshine, but low light levels and poor diet can lead to deficiencies. We find that allowing our hens to forage and adding an occasional few drops of water-soluble vitamin D3 in drinking water helps them avoid problems.

What is interior egg quality like?

Beyond individual and breed traits, each hen's eggs will vary in quality and quantity in response to dynamics of season, health, age, diet, and social status. Generally, hens thriving in ideal conditions produce eggs with richly colored, golden-domed yolks surrounded by a thick, slightly cloudy albumen (the whites). There's no need to walk on eggshells if an otherwise healthy hen lays eggs containing tiny spots of blood, double yolks, or other minor quirks. Several interior anomalies are worth mentioning, however.

Problems affecting egg quality

Clear, runny albumen Albumen gradually clarifies and becomes less firm as eggs age, providing an indication of freshness. If the eggs are known to be fresh, however, runny albumen may be a symptom of Newcastle disease or infectious bronchitis, or it could be the age of the chicken who laid the egg: older chickens sometimes produce watery eggs. The problem could also be caused by storage issues, particular a high storage temperature and low humidity situation, which speeds an egg's aging process.

Pale yolks Diet is the main factor here. Rations containing yellow corn or alfalfa meal will produce eggs with average looking yolks, but to get deeper yolk hues and firmness, ensure that your hens eat dark green leafy vegetables and colorful fruits that are rich in flavonoids, such as berries and tomatoes.

Weak yolks Carotenoids help preserve the integrity of the membrane around the yolk, and weak yolks may indicate a deficiency of this antioxidant. Carrots are a good source of carotenoids, but green algae contains an even more potent form called canthaxanthin. Supplementing with *Nigella sativa* (black cumin) can also help improve yolks.

Are there any signs of parasites?

Hens often live with a light load of internal or external parasites without this affecting egg production. A moderate to heavy infestation, however, will cause hens to stop laying or slow their output noticeably. After treatment, hens should resume laying, but their output may be slowed and egg size may be reduced for a while until they fully recover.

Recommendations Identify and treat parasites following our recommendations. None of our recommended treatments require a withdrawal period, but if the medication you use recommends this, do not eat eggs produced by the hens during this time.

Are there any signs of disease?
Are the birds overcrowded?

Too many chickens in a small area, disease, and other sources of stress can all affect egg production and shell quality.

Recommendations Maintain chicken health and follow the good care practices that we describe. Healthy chickens look sleek, clean, and bright.

CASE CLOSED

As you've probably gathered by now, our eggless coop could have resulted from anything and everything—from laying variability of our heritage breeds, to low spring light levels, to too much shade from the tree. In this case, however, it turned out that a well-meaning neighbor had been feeding our hens stale bread and cookies every day for a week. The chickens had gobbled up the treats, leaving no trace for us to find later.

RESPIRATORY ILLNESSES

You may on occasion notice one or more of your birds sneezing, suffering from nasal discharge, or having visibly swollen sinuses. In severe cases of respiratory illness, a hen will also be lethargic, her breathing may be audible, she could look puffed up, and she may lose interest in eating.

"Who coughed?"

AVIAN INFLUENZA

This highly contagious disease is closely monitored worldwide as both a livestock and potential human health threat. Mild forms are relatively common and resemble other respiratory illnesses. The highly pathogenic, scary kind is distinguished in chickens by alarming facial swelling, blue comb and wattles, dehydration, respiratory distress, dark red and white spots on legs and combs, and sometimes bloody discharge from nostrils. If you suspect a case, immediately contact both a local vet and the health department in your area to report the potential emergency—and isolate your birds. Luckily the dangerous form of Avian influenza is extremely rare.

MYCOPLASMOSIS

A common cause of respiratory illness in chickens, mycoplasmosis is a bacterial infection that results in severe symptoms including nasal discharge, breathing problems, swollen face, and swollen joints, and your hen will look genuinely ill. The problem with mycoplasmosis is that if an infected chicken survives, this chronic condition will tend to return again and again, making infected chickens gradually weaker and at risk from dying when they have to fight another illness. Mycoplasmosis can be disastrous for a backyard flock because it's contagious—if you introduce new, healthy birds into an infected flock, they can also contract the disease. Because you are likely to be keeping chickens in perpetuity, replacing those you lose over the years with young birds only perpetuates mycoplasmosis, which is resistant even to antibiotic treatment.

If you suspect a mycoplasmosis infection, follow our recommendations for nonspecific respiratory disease, but consider adding a nasal application of colloidal silver, which has good data supporting its effectiveness as an antimicrobial in animals (although it has not been studied in chickens specifically as far as we know). There is no definitive cure for mycoplasmosis, but supportive care may reduce its impact on your chickens' long-term health.

Truly, prevention is the best cure when it comes to this disease—and, most importantly, be careful about where you get your birds. A healthy flock can be infected after chickens from another flock are added. Vet your chickens thoroughly (pun intended) by asking the previous owner if the birds are generally healthy and have ever had colds or other health problems.

NONSPECIFIC RESPIRATORY DISEASE

If your hen—and perhaps some of her flockmates—is sneezing, but is otherwise doing business as usual, she probably has nonspecific respiratory condition. This is the chicken equivalent of the common cold and, happily, no cause for serious concern. It's a good time, however, to make sure you are giving your chickens probiotics and aloe juice. Watch them closely for signs of more serious illness until they seem to get better. Many chicken keepers report good results after treating nonspecific respiratory ailments with a product called VetRx Poultry Remedy, a natural solution made from Canada balsam, camphor, oregano oil, and rosemary oil. You dissolve it in warm water and administer it nasally or orally. We think that the oregano and rosemary oil, which are powerful antimicrobials, are behind the good results. You can also mix up and use homemade antimicrobial herbal oil spray, because it includes more botanical ingredients with proven beneficial effects in poultry. RopaPoultry Complete Oregano Oil+ Supplement can also be effective.

LESS COMMON RESPIRATORY AILMENTS

Several chicken respiratory diseases are more severe than mycoplasmosis, but they are thankfully not common in backyard flocks. Because these are serious business and require the services of a veterinarian—and possible euthanasia for the entire flock—we keep treatment descriptions brief and group them together here. Fowl cholera, infectious coryza, Newcastle disease, aspergillosis, and infectious bronchitis all show respiratory symptoms similar to those we've described, but additional worrisome symptoms will alert you to the fact that something more serious is affecting your hens, such as dramatically swollen faces, gasping, abnormal egg development, or an alarmingly quick death. Call your vet immediately for a consultation if you suspect a serious condition. Early intervention may save some of your flock, depending upon the diagnosis.

Don't try to diagnose a serious respiratory illness yourself. Time without appropriate treatment only reduces the likelihood that a hen can be saved, and in the worst-case scenario, it prolongs her suffering before she can be euthanized. Of course, some folks just let nature take its course and allow a chicken that is destined to die go naturally. The argument against this, however, is that if a hen survives, she will likely be sickly for the rest of her life, will often be a poor layer, and may also be contagious, leading to other sick hens. A veterinary consultation will tell you which respiratory disease you are dealing with, whether survivors will be carriers of the disease, and how long the coop and run should be considered contaminated before you can introduce a new flock.

If you do not or cannot get a hen to a vet for treatment before you lose her, or more birds, consider sending in one of your lost birds for a necropsy. Usually, your state agricultural extension service can provide this service at some cost. Also try contacting the nearest university that offers a veterinary program, which may be able to provide the information you need.

BIOSECURITY BASICS

Quarantining new hens and observing their overall condition for several days is a basic strategy to prevent the introduction of disease to your flock. You should also dedicate a pair (or several pairs) of rubber boots to use exclusively in your own coop, and never allow others who own chickens to enter the coop or run until they have slipped on a pair or donned disposable booties over their shoes. The converse is equally important when you visit other people's coops. It's universally recommended that you not allow contact between your flock and wild birds; although we understand the reasoning behind this, we find it to be impractical in most cases. If you suspect a wild bird is sick or you find a dead bird near your flock, don't hesitate to contact a local animal welfare agency. They will be very interested, mostly because of the implications for human illness outbreaks. For the same reason, be alert for any fast-moving, devastating illness in your own flock and follow our guidelines to treat them, or seek out a vet's opinion if you're unsure what to do.

WOUNDS AND INFECTIONS

·· »

Opposite: A bubbly discharge from the nostrils and a swollen face are associated with several of the more severe respiratory illnesses.

By far, predators are the most common cause of serious chicken wounds. This is usually preventable if you build an impenetrable coop and run. Despite precautions, however, we have had our share of unanticipated predator attacks—but it's always been our fault. Overconfidence, miscommunication ("I thought *you* were going to close up the coop!"), or mechanical failures are involved.

Even if an attack occurs because of an automatic door failure, it's still the keeper's fault: remember that any power outages, whether caused by unplugging the system or by other factors, will make the door's timing wonky. All self-flagellation aside, a predator attack often means you will have one or two dead chickens and an additional wounded warrior on your hands. In your grief and self-recrimination, you will want to do all you can for your injured ladies.

PREDATOR WOUNDS ··

In treating an open wound, your goals are to prevent excessive blood loss, manage shock, and prevent infection of the wound while it heals. You will likely find a wounded chicken trying to hide somewhere, huddled up, and so stunned that she is easy to catch. Gently gather her up and separate her from the rest of the flock. Bring her to a dark, quiet area of your home for the first twenty-four to forty-eight hours, at least, or until you notice that some of her spunkiness is coming back. When your injured hen has perked up and rejoins her flockmates, you may need to cover the wound to keep her friends from pecking at it.

In this protected, quiet environment, you can attend to her wounds. Your first priority is to clean out the wounds. We keep Vetericyn or iodine on hand for this purpose. Saline solution is another option for flushing a wound if you think you need a large volume of liquid to wash away debris. Single-use syringes without needles are perfect for this purpose. Flush the wound until there is no more visible grit or grime.

Often, the wound will be on the chicken's body rather than on a limb or her head, but if you find a wound on a limb, feel for evidence of a broken bone. If a wing is broken, it will be hanging limply rather than being tucked up in a normal manner along the chicken's side. A leg or toe break is often obvious because of a strange angle to the bone, limping, or tenderness to the touch. If you believe she has a bone break, do your best to splint the bone using a short portion of dowel or a finger splint attached with duct tape on either side of the break to hold the bone more or less in the correct position as it heals. We keep dark-colored duct tape on hand for chicken wounds—anything bright or shiny like the classic silver-colored duct tape will encourage flockmates to peck at the area.

After cleaning the wound, spray it with colloidal silver, and then apply some homemade antimicrobial salve. If do-it-yourselfing is not your style, a good, natural, ready-made salve for animals, GreenGoo, is available from the kind folks at Sierra Sage.

ANTIMICROBIAL WOUND SALVE

Use this wound salve to discourage bacterial colonization by creating an antimicrobial barrier. Petroleum jelly provides an inexpensive and effective barrier to germs, and it's a byproduct of an industry that is not going away anytime soon. A more expensive alternative is Waxeline ointment, a product that contains no petroleum. Adding the gentian violet to the mixture will stain your bare hands and chicken skin and feathers blue, so wear disposable gloves when you handle it. But it can be helpful for a wounded chicken, because the blue color masks the healing wound's redness, which would otherwise entice other hens to peck at and explore the wound.

½ cup petroleum jelly or similar ointment base

2 tablespoons manuka honey

1 teaspoon aged garlic solution

1 drop white thyme essential oil

(optional) ¼ teaspoon gentian violet

In a lidded container, mix ingredients together. Clip the feathers in the area flush to the skin if possible and apply as needed.

When bandaging a severe wound, try to pull together the edges of skin around the wound to help close it, and use dark-colored duct tape to hold the wound edges together. Letting a wound heal by enabling the skin edges to knit themselves back together is the best way to promote quick healing. (We haven't had difficulty removing the duct tape later because our hens are such vigorous dust-bathers that the tape is usually pretty loose by the time we remove it. If you are concerned about harming a hen's skin, and the tape seems firmly attached, rubbing some mineral oil under the bandage can help.)

If, as is often the case, there appears to be a large portion of skin and perhaps muscle missing, it will not do your patient any good to try to pull the skin together; you don't want to add tension to the severely wounded skin. In this case, the wound will need to heal from the deeper layers up to the surface. This is a much slower healing process, but it's absolutely necessary if a significant area of skin or tissue is missing. Because it does take a while to heal, you need to be especially vigilant about preventing infection until the wound is healed over with a new layer of skin.

Believe it or not, the skin will grow back. And then, on top of that, new feathers will develop.

If the wound is positioned such that bandaging it well seems impossible, use your ointment with gentian violet or spray it generously with Dr. Naylor Blu-Kote (with active ingredients alcohol and gentian violet). Alcohol stings like nobody's business on a wound, and that is our main concern about Blu-Kote. It has not been formally tested on chickens, but we know lots of chicken keepers who use it successfully.

PECKING WOUNDS

There is a completely logical etymology for the term "pecking order." It is entirely normal for chickens to use a bit of pecking, and what appears to us to be bullying behavior, to determine the social rank in a flock, even a small one. For us, it is nearly as painful to watch as when our children bicker over something inconsequential. We want to stop it, break it up, and throw up our hands and scream, "Can't everyone just get along?" But they have to work it out themselves to establish a functional social order.

Roxie the Third stands boldly in front of her chicken swing, which keeps her busy and out of trouble.

To prevent pecking wounds in the first place, give your chickens enough space to roam and use boredom busters to keep them busy. Boredom and overcrowding can contribute to pecking behaviors. Even in the best possible conditions, however, an overzealous hen will occasionally chase, peck, and generally intimidate a chicken that she sees as posing a threat to her total domination, and occasionally it can get bloody. A pecking wound usually heals with little incident if you can keep it from being pecked at repeatedly. If the wound is a scratch, not actively bleeding, and has no sign of infection such as foul-smelling discharge, put some Blu-Kote or gentian violet on it. Yes, your chicken's feathers may be blue until her next molting, but it's worth it compared to the alternative. See page 56 for more advice on handling aggressive chickens.

BUMBLEFOOT

This aptly named malady is an easily identifiable infection of the chicken's foot. You will become aware of it because an infected chicken will be limping. She will also be easier to catch than usual, and when you pick her up to get a closer look, you'll notice a swollen lump on the bottom of her foot, often with a black discoloration in the center of the swollen area. The infection is caused by *Staphylococcus aureus*—a staph infection.

The most common cause of foot injury is an inappropriate perch with rough or sharp edges that cut into a hen's foot. However, any break in the skin of the chicken's foot, even caused by an errant splinter or bit of chicken wire, can let bacteria in to cause an infection.

If there is distinct swelling, but no black spot or scabby appearing area, you may be able to treat the infection topically. Epsom salt baths will encourage the infection to come to a head and drain. Apply a small gauze pad infused with homemade wound salve, and hold it in place with dark-colored duct tape. Give this a couple of days to work and then check out the swelling. If it appears improved, continue with this regimen, replacing the dressing every two or three days until the area looks normal again.

If, after treatment, symptoms worsen, you need move on with draining the infection to avoid progression to sepsis (a systemic and life-threatening infection). The swollen area will likely be full of pus, or, if it is a long-standing infection, there may be a hardened pus plug in the wound. Call your veterinarian if dealing with this is not your thing. On the other hand, if you take a macabre sort of satisfaction in delving in and eradicating grossness, know that it's something you can do yourself.

Start with another Epsom salt bath to clean and soften the skin in the area. The longer the soak, the better for this, but don't let the water get cold. Remove her from the water and wrap her snugly in a towel, with only her wounded foot and head out. If the edge of the black area on her foot is crusty like the edge of a scab, you may be able to grab the edge of it with a set of tweezers and uncap it.

If this does not work, you can make a small incision in her foot—with the help of an assistant. If you have a scalpel, by all means use it. If not, sterilize a sharp knife (such as a craft knife or paring knife) by dipping it in a diluted solution of 1 cup water and 1 teaspoon bleach, or 1 cup vinegar and 1 ½ teaspoons of 3 percent hydrogen peroxide. Leave the knife in the disinfectant for about ten minutes before using it to ensure complete sterilization.

As you make the incision, repeat to yourself, "I am helping my chicken. I am helping my chicken. I am helping my chicken," which will steel your nerves. The incision needs to be large enough to allow the abscess to drain. The size of the incision will depend on the severity of the bumblefoot infection, but usually a tiny incision of about 1.0 cm long and 0.5 cm deep will be sufficient. There is usually surprisingly little blood involved.

Like any abscess, it is essentially a very large pimple. After you have opened up the abscess, apply gentle pressure down around the edges of the swelling to encourage the pus to drain. The swelling should be considerably diminished, but not entirely gone. Take a good look at the foot to make sure you've drained the wound completely. The surrounding tissue will be swollen, with no liquid or semi-solidified pus. Occasionally, you may need to grab your trusty tweezers again and go after a pus clump. Again, remind yourself that this will make your chicken feel so much better.

Once you have the wound cleared out and the tissue inside looks fairly clean and pink, cover it with homemade salve-soaked gauze and dark-colored duct tape.

Replace this every two or three days. You should see improvement every time you change the dressing. You can stop bandaging the foot when it looks like the skin is intact, after two or three weeks.

After allowing yourself the pleasure of having taken care of something that is truly gross, make sure that the chickens' perch is rounded, that no sharp debris is lying around, and that none of your other ladies are similarly afflicted.

NEUROLOGICAL CONDITIONS

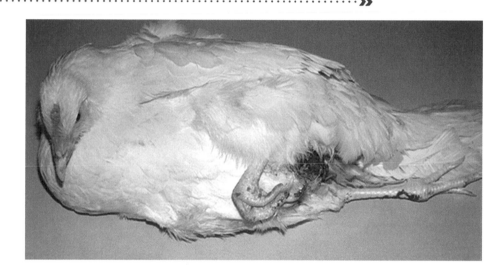

This poor pullet shows the paralyzed posture classic for Marek's disease.

Chickens may suffer from a variety of diseases affecting the nervous system. Though disease may be present from hatch, most congenitally deformed or diseased chicks never make it past the hatchery doors. More commonly, illnesses occur later, resulting from underlying nutritional deficiencies, pathogens, or various toxins in food or environment. Each neurological disorder is unique, but each is generally marked by symptoms such as paralysis (localized or general), difficulty breathing, erratic behavior, and contortions of limbs or neck.

MAREK'S DISEASE

Marek's disease is the most serious neurological disease that affects chickens. It is caused by a herpes virus that results in paralysis, tumor development, and very often death in unvaccinated chickens. It is highly contagious and spreads through skin and feather dust. Some birds will survive Marek's, but even vaccinated birds (who are much more likely to live) will become infected with the virus if exposed and will remain highly contagious for life. If an infected bird is moved into a flock with unvaccinated birds, the other birds will become infected as well, and the ground in their coop and run will be contaminated for years.

The classic sign of Marek's is paralysis, in which one leg is thrust forward and the other backward in a posture that lasts several days. In the afflicted hen's eye, the shape of the iris may also become irregular, with the pupil abnormally dilated or constricted. The hen might eventually die from tumor development. In a commercial setting, these birds are euthanized. In nature, they die from inability to reach food or water. If a backyard chicken keeper were to hand-feed and hydrate a chicken with Marek's, the hen might improve and even survive, but she would likely suffer from ongoing health problems, and she very well may pass on the virus to other birds.

We strongly recommend buying only Marek's vaccinated chicks and making sure that any adult birds you acquire have also been vaccinated. Unfortunately, the hatchery we use most often tells us that vaccination is the exception rather than the rule, and we think this is probably the case at other hatcheries as well.

There is a ferocious vaccine debate in the chicken-keeping community: to vaccinate or not? Marek's vaccine contains live attenuated virus. Attenuation takes an infectious agent and alters it so that it becomes harmless or less virulent. The form of the virus in the vaccine, therefore, should not cause severe illness but will trigger an immune response that should provide protection if the vaccinated chicken encounters the nonattenuated virus. Some folks worry that if some of their chickens are Marek's vaccinated and others are not, the vaccinated chickens could pass the disease to the nonvaccinated birds. But if they did, it would be the attenuated version. Conversely, other folks figure that if they have at least some vaccinated birds, the attenuated virus will get passed around and protect the rest of the flock. But this would not offer them the degree of protection provided by actually getting the birds vaccinated.

The most serious concern is that if a vaccinated bird survived after exposure to Marek's, that surviving bird could then be contagious and prolong the survival of a particularly nasty strain. In the past, before vaccinations were available, virulent strains of Marek's disease tended to die out because infected birds all died. These days, however, most strains are not the super-virulent variety, and vaccination could enable all of your chickens to be unaffected by milder strains. That seems like a good thing to us.

If Marek's disease sweeps through your flock, your responsibility would be to avoid keeping chickens until you have been advised by a veterinarian that it is safe to do so. The absolute worst thing you could do would be to give a Marek's survivor to another chicken keeper. (Similarly, it is never a good idea to visit a yard in which all the chickens have recently died, or to bring a lone survivor home to your flock.) The really important lesson here is that if you ever have a chicken that shows these symptoms, you no longer have the option of bringing unvaccinated chickens onto your property. The virus can remain in the soil for years, with or without the presence of chickens.

When a chicken shows signs of weakness or paralysis, it is common for a chicken keeper to suspect that Marek's disease is to blame. However, especially if you're sure the chicken was vaccinated, you should consider other causes.

OTHER DISEASES
WITH NEUROLOGICAL SYMPTOMS ·······················

Several other neurological problems can affect chickens. It's a good idea to send any bird that died after exhibiting neurological symptoms to a lab for a necropsy so you can know for certain what your flock has been exposed to; this will help you create a preventative treatment plan.

✦ **Aspergillosis** This is generally considered a respiratory disease and is caused by fungal infection. The classic sign is a chicken gasping for breath, but there can be neurological symptoms if lesions develop in the brain or eyes. Blindness can occur, along with a severe contracture of the neck and paralysis. The hen's eyes are usually cloudy or affected by lesions rather than having the misshapen irises indicative of Marek's disease.

✦ **Fowl cholera** This bacterial infection usually causes sudden widespread death in chicken flocks. Survivors can show neurological symptoms. Rodents can be carriers of *Pasteurella multocida*, the bacteria that causes fowl cholera, and should be controlled. The bacteria survives without a host for only a few weeks.

✦ **Lymphoid leukosis** Birds with this tumor-causing virus can show neurological symptoms in addition to depression, lack of appetite, weakness, diarrhea, dehydration, and emaciation. Symptoms are more likely to be present after a hen reaches fourteen weeks (later than Marek's) and more typically after egg-laying is established. It is important to differentiate lymphoid leukosis from Marek's, because leukosis is not nearly as contagious and is much easier to eradicate from a site by disinfecting the area.

✦ **Newcastle disease** Generally considered a respiratory disease, Newcastle is caused by a virus that can affect the neurological system in a manner very similar to aspergillosis. There are severe strains, but none are currently known to exist in the United States.

PARASITES
·· »

Chickens are the unhappy potential victims of numerous internal and external parasites. At the risk of sounding repetitive, we believe that much of this problem can be prevented with good feeding, litter management, avoidance of overcrowding, and vetting and/or proper quarantine of new birds. A little botanical proactive management is beneficial as well.

INTERNAL PARASITES

Several types of worms can plague chickens, but roundworms, gapeworms, and tapeworms are the most common. Any new adult chicken introduced to the flock should be presumptively treated for worms during its quarantine period, and an annual deworming protocol is advisable for all your chickens, because often there are only vague clinical symptoms of an infestation until it is very heavy. Because even a light worm load can result in reduced egg production, it makes sense to keep worms under control.

Aloe is effective against internal parasites, so you are a step ahead if you are already adding it to your chickens' water every day. Adding uncalcined, food-grade diatomaceous earth at a ratio of 1 cup to every 3 pounds of feed for one month should complete the process. Raw pumpkin seeds are a popular traditional backyard chicken keeper's remedy for worms. Pumpkin seeds are a high-protein food and beneficial to general poultry health, so by all means, give them a try. Another commonly recommended backyard remedy is to add cayenne pepper to their feed—about 3 tablespoons per gallon of feed. Cayenne has been shown to be safe and well tolerated.

EXTERNAL PARASITES

Mites and lice of various types are very common pests for poultry. Ticks and fleas can also be an issue, but, happily, all are effectively treated in the same way, so you don't need to feel obligated to make an accurate diagnosis about the type of critter that's afflicting your flock to treat it.

Prevention, as always, is key. All new birds (except baby chicks) should get a neem lice and mite drench upon arrival and then a dusting with uncalcined, food-grade diatomaceous earth every few days for the length of their quarantine. Tell your hen that she is getting a show-bird treatment to make her feel special. If you see any floating insects in the water when you are done, you know for sure that you need to follow through with the diatomaceous earth.

Also keep an eye on your chickens' legs. The scales should always lie smoothly together. If the scales seem to be pulled apart or you see gaps between scales, you are likely looking at a scaly leg mite infestation. Rub a mixture of three parts coconut oil and one part neem oil into the legs, filling all the spaces between the scales. Repeat this daily for a week, and then continue to apply it weekly until the scales return to normal. Keep applying the ointment for at least thirty days to make sure you have covered an entire mite life cycle.

To keep mites at bay, avoid using straw inside your coop; it is a favored hiding place for red mites that lurk in the cracks and corners of the coop to feed on your poor chickens at night.

Hens depend on dust baths and sunbathing to help keep mites and other parasites at bay. You can build them a dust-bathing box outside the coop and fill it with two parts coarse sand and one part uncalcined, food-grade diatomaceous earth. Or, if your ladies already have an established dust-bathing area and it has become a good-sized pit, top it off with diatomaceous earth. Make sure the dust bath is in a dry and protected area so it can be functional even in wet weather.

Opposite, clockwise from top left: These bedraggled chickens appear to be molting, but their feet show signs of leg mites, and their general appearance is suspicious for an infestation.

A healthy chicken's foot should have smooth scales and show no swelling or redness.

The scale gaps and discoloration of these hens' legs indicate a severe case of scaly leg mites.

The rough scales on this hen's feet and legs are telltale signs of scaly leg mites.

Finally, if you ever see your girls grooming in an agitated, uncomfortable, obsessive manner, this should strike you as very different from their usual contented grooming and give you reason enough for an exam. If a hen has tattered feathers, or you see tiny insect eggs at the base of her feathers or irritated skin, especially around her vent or under her wings, you need to treat for mites. Give her a neem lice and mite drench, and then rub in a mixture of three parts coconut oil and one part neem oil directly onto the irritated skin or areas with signs of mites or eggs. Blow-dry her feathers and dust her thoroughly with food-grade diatomaceous earth. Repeat the dusting every two or three days for at least ten days.

FLIES

Most chickens produce enough droppings to support a moderately large population of these buzzing beasties throughout the summer months. From time to time, the fly population will spike in response to changes in weather, a buildup of waste, or other causes. Although you might feel compelled to resist flies vigorously, keep in mind that complete eradication is impossible.

Several species of flies commonly hang around coops in North America, and although most are merely unpleasant visitors, some can spread disease and injure your hens.

Opposite: Chickens instinctively dig pits for dust bathing, which is helpful in controlling parasites—though it might take a toll on your garden.

Opposite, bottom left: A chicken dust bath in action

Opposite, bottom right: A tiny white cluster of eggs at the base of the feather provides an obvious diagnosis of poultry lice.

HOMEMADE COOP SPRAY

Makes enough to coat the interior surfaces of a 4-by-4-by-4-foot coop

½ cup Murphy's Oil Soap

½ cup warm water

2 tablespoons neem oil

5 drops thyme oil

2 drops clove oil

2 drops lavender oil

2 drops rosemary oil

2 drops sage oil

Mix these ingredients in a spray bottle. Spray the mixture inside the coop after cleaning on a warm, sunny day, so it can dry by the time your chickens are ready go back in for the night. Make sure to spray the cracks and corners. After the coop is dry and less pungent smelling, sprinkle uncalcined, food-grade diatomaceous earth in all the edges and cracks of the coop. Finally, add your litter of choice.

Black soldier flies These flies are actually not problematic. A non-biting native insect, *Hermetia illucens* is not known to carry disease. They are, in fact, considered beneficial insects, because their larvae are predators of other fly larvae and they consume manure and food waste very quickly. Some chicken keepers even raise soldier flies to produce high-protein grubs for their hens.

Blowflies These really bad guys, *Calliphora vicina* (also known as bottleflies), are iridescent green flies that lay their eggs on flesh at wound sites. The hatching larvae feed on the flesh, leading to infection and more yummy dead tissue for them to eat. If there's enough to eat, a minor infestation can rapidly progress into a serious, potentially fatal condition known as fly strike. Always be on the lookout for wounds and dirty bottoms on your hens, and note any changes in their bodily odor, weight, or energy levels. If you notice any of these things, it is time for an inspection (wear disposable gloves). Look first around the vent for signs of maggots or a wound and note any odor. Continue your inspection by gently moving feathers aside to see the skin underneath, paying particular attention to back, tail, and underbelly.

If you see larvae, it's a good idea to take your hen to the vet immediately. If you want to try to work on a minor case yourself, begin by removing the maggots with your gloved hands and putting them in a plastic bag; then tie it closed and toss it into the garbage. Clean the wound with a gentle saline wash, and give the hen a bath if possible to clean her feathers. After she's thoroughly dry, treat the wound using our recommended methods for general wound care. Do not put the hen back outside until her healing is well underway, which can take several days. A cardboard box, pet carrier, or dog crate will work well to house her indoors.

Keep in mind that none of this can happen unless there is an untended, untreated wound in the first place. We think this particular horrific scenario provides a very strong argument for keeping a close eye on your girls.

The black soldier fly is ⅛ to ¾ inch long.

False stable flies *Muscina stabulans* is common around poultry. Adult flies don't bite, but they can spread disease via bodily contact. They're not all bad, though—near maturity, the fly's larvae can feed on housefly larvae. You should be moderately concerned if you notice them around your hens.

Horse flies *Tabanus abactor* are less commonly found around poultry than near larger animals such as cows. They are big, nasty, biting, disease carriers that should be targeted if you see them.

The blue blowfly, a real bad guy, is 10–11 mm long.

The false stable fly is about 8 mm long.

The horse fly is about ½ inch long.

The common housefly is about 7 mm long.

Houseflies Common indoor pests *Musca domestica* may occasionally show up where chickens live. They are not typically disease carriers, but their presence in large numbers should be considered a call to action.

END-OF-LIFE CONSIDERATIONS

Most chicken keepers who maintain pet hens become emotionally attached to them and feel naturally compelled to nurture and care for them. Similarly, those who keep small flocks for production purposes usually have an ethos strongly rooted in being good stewards and caretakers of their flocks and try to minimize or reduce suffering. With these aims in mind, we feed carefully, maintain a healthy environment, and attempt to heal our sick and injured birds. How we handle an illness or injury that can never be cured or will never heal properly, or humanely slaughter a chicken destined for the table, is an important topic.

Over the years, we've seen a number of chickens pass away from both natural causes and predation. When a hen dies from illness, we somberly evaluate what happened, do some research to discover the root cause, and resolve to respond more effectively in the future. When our hens have been lost to a violent attack, our emotions are raw and close to the surface, but in the end, we shore up our defenses and try to remember the good times.

The most difficult experiences have occurred when a hen has been severely injured or is clearly suffering from an illness from which she will not recover. With our bellies in knots and lumps in our throats, we must decide whether it's more humane to put her down or continue to administer care and make her as cozy as we can. We must admit that we have yet to put down an ailing hen that was obviously suffering from a terminal illness, instead taking the slightly cowardly "treat

FLY PREVENTION *and* REMOVAL

Neighborhood dogs and cats, yard debris bins, garbage cans, and living and dead wildlife can all contribute to a local fly population. Urban chickens will inevitably contribute their share, but you can avoid fly population spikes by taking a few prevention measures.

Good litter management is the cornerstone of fly control. Continue to add fresh litter, maintain a low pH, add beneficial microbes, and keep things relatively dry, and fly problems both inside the coop and outside in the run will be largely avoided. However, sometimes flies can be persistent, and you'll need the help of some practical remedies.

➔ **Baited liquid traps** Our favorites, these traps are satisfyingly effective. Various baits lure flies into a bag or another container, where they drown and accumulate into a satisfying mass of thousands of dead bodies. Be warned—the bait and the dead flies stink, so you'll want to place them near the coop but away from neighbors and outdoor living spaces.

➔ **Diatomaceous earth** The uncalcined, food-grade type is useful for combatting and possibly preventing fly infestations. Sprinkle it onto the litter and add it to your hens' dust bath areas.

➔ **Sticky tape** This is a classic fly remedy, but we have been accidentally stuck to a dangling roll covered in fly corpses too often to recommend it. Fly wrapping tape is a more effective modern product; the long reel of tape can be strung high in coops and barns where flies land and rest.

and make comfortable" approach. On the morning when we inevitably find that she has died, we feel both sadness for her loss and a bit of guilt at our indecisive handling of the situation.

There are other options, however, that are probably better than this. A quick and efficient killing by an experienced hand is likely the solution that causes the least suffering. The option that we recommend, however, is to take your hen to a vet as you would any other pet. This will enable you to confirm diagnosis and obtain valuable advice, and you can be assured of humane euthanasia for your hen. If you have previously established care with a vet, you should have some idea of the cost of euthanasia.

Whichever technique you opt for, your priority must be reduction of stress and fear—for both you and the bird. Begin by taking your hen someplace quiet and far enough away from the rest of the flock that no one else will become alarmed by the event. Hold her gently under your arm, share bodily warmth, and softly compress her wings. Stroke the underside of her neck and belly, not the back of the head if possible, and speak in a soothing voice.

Please consider skipping this information if you are distressed by the topic of chicken euthanasia.

Techniques for ending a life

+ **Anesthetics** These can be administered, preferably by a veterinarian, as a prelude to another method, or as an overdose to induce death.

+ **Cervical dislocation** This is a common practice on small farms, where it's referred to as "wringing their necks." Some research indicates that consciousness is sometime lost only slowly, however, and that the hen may suffer, particularly if you lack the experience and confidence to act decisively and without hesitation. This advice also applies to decapitation, which is fast and effective when done correctly but prone to bungling by nervous or inexperienced owners, and it can lead to human injury as well.

+ **Inhalation agents** These substances can be administered by veterinarians and laboratory technicians; they're able to administer a controlled dosage in a special chamber. Amateurs may have successfully simulated these techniques using car exhaust or inhalants, but experts warn that most of these methods may cause needless suffering and in some cases can harm the person administering it. We recommend that you steer clear of trying this method at home.

+ **Lacerating the carotid arteries** A time-honored, traditional method, it's the only one we've personally performed. As so-called chicken experts, we long knew that it was our sacred responsibility to perform the killing act that we had described for our readers and customers, and we decided to do so with two heritage roosters gifted to us for meat.

  Place the chicken in a restraining cone (also called a killing cone)—it's sort of like an upside-down traffic cone mounted to a wall. Place a bucket on the ground underneath. The bird should calm down quickly once it's placed head first in the cone and may seem to fall asleep. Gently take hold of the beak and elongate the neck slightly with one hand, without disturbing the bird's trance-like state. Hold a very sharp and straight, nick-free knife in the other hand. Make one confident cut across the full front of the neck just below the jaw, applying firm pressure to ensure that you sever both carotid arteries simultaneously. Blood will come slowly at first, but it should steadily collect in the bucket below—it amounts only to a couple of ounces, but you should let it bleed out for about a minute and a half to ensure that it all drains out. The bird should seem to lose consciousness immediately, but there may be a few leg spasms as it loses neurological function; these are normal but can be disturbing. Full unconsciousness is achieved when the bird ceases to have a blink reflex when the cornea is touched.

BLENDING *Your Own* FEED

Mixing your own poultry feed is a common practice for subsistence farmers and permaculture devotees around the globe. It's also hugely satisfying and achievable for the highly motivated chicken keeper seeking a more personalized, regionally grown, and healthful alternative to premixed feeds.

With little effort, our garden has made a meaningful and frugal addition to the energy content of our feed. We've used a combination of homegrown corn (fresh and dried), greens, potatoes, stale bread and cereal, and a variety of fresh and frozen fruits in concert with a concentrated, high protein feed to improvise a complete ration. We appreciate the sustainable nature of this approach, but the hens seem more pleased about the variety of textures and tastes this adds to their primary diet of pelleted feed.

If you are primarily motivated by thrift, keep in mind that your from-scratch feed will generate savings only if you are able to use a substantial portion of homegrown ingredients or you buy in bulk—and in this case, bulk means whole bags, not scoops from the bulk bins at the grocers. Mixing your own feed in this way is most practical for large flocks, because to realize any savings whatsoever, you'll need to purchase each of your major ingredients in a minimum of 50-pound bags and full pallets of 2000 pounds each.

Here's the math: In our area, bagged feeds costs between 45 cents and 95 cents a pound. Use this for comparison when shopping for your components. A quick scan of a natural grocer's bulk bins will reveal the challenge: organic grains and legumes are expensive, around $1.95 to $3.95 per pound. At these prices, a 50-pound bag of home-mixed feeds would cost about $100, not including relatively expensive optional supplements and essential vitamins and minerals and enzymes!

CONCENTRATES

It was once common for farmers to feed their flocks with rations made with purchased feed concentrates blended with bulkier ingredients grown inexpensively on the farm. Concentrates are nutritionally dense, containing little or no energy (grain) but plenty of protein. They typically include small amounts of other ingredients impossible for farmers to manufacture themselves, such as vitamins and minerals, probiotics, and enzymes. The genius of this approach is that the energy (grain) portion of poultry feed is considerably bulkier and more expensive to transport, but it's cheaper to produce, whereas, as the name suggests, concentrates are compact, delivering more nutrition in a smaller volume. By growing the grains and transporting only the protein and minor ingredients, farmers have the best of both.

We're currently developing our own version of a concentrate that urban flock feeders can use to balance an economical, local source of energy. For instance, our concentrate would enable you to grow a supply of cultivated corn, potatoes, or other high-energy crops appropriate for your area and blend it with more difficult to produce, specialized ingredients needed to make a nutritionally complete food for your flock.

If you want to get really thrifty, you could combine our concentrate with a locally available, high-energy, edible resource that's currently being composted or discarded. For example, you could partner with a bakery to take stale but edible bread to feed your hens. Assuming that the bread contains about 11 percent protein, and your concentrate contains 40 percent protein, you could use a tool such as Pearson's Square to calculate the ratio of concentrate needed to balance the bread to achieve your particular feeding goal.

Pearson's Square is a simple, quick way to calculate the relative weights of two feed components needed to meet a particular nutrient requirement of your flock. It's most effective with two ingredients, but it can also be used to balance more than two if you are better at math than we are.

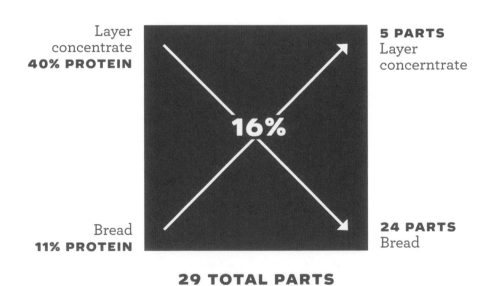

Layer concentrate **40% PROTEIN**

5 PARTS Layer concerntrate

16%

Bread **11% PROTEIN**

24 PARTS Bread

29 TOTAL PARTS

Pearson's method for two ingredients

1. Draw a square. In the center of the square, write the percentage of a particular nutrient, typically protein, that you want to achieve in your mix. In our example, we're using bread and layer concentrate layer ration, so write "16%."

2. Now draw diagonal arrows, as shown, from the upper left corner to the lower right corner and from the lower left corner to the upper right corner.

3. At the upper left corner, write the name of the first ingredient (layer concentrate) and its content of the nutrient you are balancing (protein in this example) expressed as a percentage (40%). At the upper right corner, write "parts" and the name of the same ingredient, but leave a blank space and no percentage (____ parts layer concentrate).

4. At the lower left corner, write the name of the other ingredient (in this case, bread) and its content of the nutrient (11 percent). At the lower right corner, write "parts" and the name of the same ingredient, but leave a blank space and no percentage (____ parts bread)

5. Now subtract the center number from each of the corner numbers, following the direction of the arrows, and write this value in the blank lines in each corner. Be sure to follow each arrow, and ignore the negative numbers (because only the difference between the two numbers is important).

6. In our example, we subtract the middle number from the upper left (40 – 16), follow the arrow, and we can write "24 parts bread" in the lower right corner. Repeat from lower left to upper right: 11 – 16 = –5, but we ignore the negative number, and write "5 parts layer concentrate" in the upper right corner.

This tells us that to achieve 16 percent protein in our mix, we need 24 parts bread to 5 parts layer concentrate. You can now mix 24 units (ounces, pounds, grams, kilograms, or other unit of weight) of bread with 5 of the same units of the layer concentrate, mix well, and you'll know that you've got 29 units of 16 percent protein homemade layer mix.

(Note that it's a little more complicated than this because the ingredients are supposed to be alike—both dry in this case—and some nutrients other than protein in the bread are duplicated in the concentrate, meaning there will be an excess. To be more precise, you'd need to dry the bread first, but there's not much to be done about the nutrient duplication—it should do no harm.)

We should say right up front that we are not as skilled as agronomists or professional animal nutritionists employed by mills to make feed formulas, and we do not have access to the tools and detailed ingredient analysis they use for this often complicated process. In our experience, this example feed recipe will get the job done, but if you are determined to feed your hens more precisely, you should consult with a qualified specialist from an agricultural university or mill.

A seemingly endless combination of ingredients can be included in a ration. This one features a few ingredients that you may be able to grow yourself or obtain inexpensively. Note that your results will be affected by many variables, including the specific ingredients you use, other things you feed your chickens, and the mistakes that authors of chicken books make.

SUNFLOWER *and* CORN LAYER FEED MIX

This is a basic 17 percent protein homegrown feed mix for adult laying hens.

6 units dried cracked corn

1 unit hulled and roasted sunflower seeds

1 unit flax seeds

½ unit Fertrell Poultry Nutri-Balancer (vitamins, minerals, and kelp)

¼ unit brewer's yeast

⅛ unit limestone grit (plus free choice oyster shell)

HOW *to* MAKE BOKASHI

Bokashi is both a thing and a process. The thing bokashi is created using EM/BM, a microbial inoculant that has been activated by growing it onto bran. When activated bokashi bran is added to suitable bulk materials (such as food waste and animal bedding) in the relative absence of air, fermentation occurs, and in the process, it acidifies the material and renders it inhospitable to pathogens such as *E. coli* and salmonella. Better yet, the fermentation results in the proliferation of healthful microbes such as those found in yogurt.

You can purchase bokashi ready to use or make your own. To make bokashi, you'll need a large bucket or another airtight container with a lid and a spigot, in which you add a blend of beneficial microbes to a grain bran substrate, fresh water, and molasses. The spigot is needed to drain the liquid that accumulates as the material decomposes.

BOKASHI

Makes 10 pounds

10 cups water

4 tablespoons molasses

4 tablespoons EM/BM

10 pounds bran

Mix the water with the molasses to dissolve the molasses. Mix in the EM/BM. Then add the liquid to the bran and mix thoroughly. Squeeze some of the bran into a ball. It should hold its shape with no extra liquid. If it's too dry, add a little more water. If it's too wet, add a little more bran. When it's just right, put the mixture into a sealed container, press it down to compact it, and place the container in a warm location. Let it ferment for at least two weeks.

When fermentation is complete, you may see white mold on the bokashi; this is good. If you see green or black mold, it means air got into the container or the mixture was too moist, and it should be discarded. Bokashi needs to ferment for at least two weeks and can then be dried for long-term storage (up to several years in a sealed container).

SAMPLE HOA LETTER

Here's a sample letter you can personalize to send to a homeowners' association to advocate for a policy change in favor of chicken keeping in your neighborhood.

To Whom It May Concern,

I am writing today to urge you to change the regulations about keeping chickens within our neighborhood. Urban and suburban chickens have many positive benefits for families and communities. I am aware that chickens carry a connotation of having potential negative impacts on a neighborhood, but by employing good practices, chicken keepers around the country routinely engage in this rewarding hobby while coexisting harmoniously with their neighbors. In fact, towns and neighborhoods in our area, including [*do a little research and name one or two nearby locales that have chicken-favorable laws here*], allow homeowners to keep flocks of hens in environments similar to ours.

Responsible adults are easily capable of keeping two or three hens in urban and suburban environments without imposing negative impacts on their neighbors. A perception exists that chickens are noisy, but only roosters crow, and they are not allowed within the [*research to make sure this is true for your locality, and then insert your city name here*] limits anyway, and we would not be keeping roosters in our yards. Cared for in the usual manner, a small flock of hens will produce no odor, are rarely heard beyond 20 to 30 feet from the coop, and produce zero visual impact when kept within the backyard or behind fencing.

I have found that the regulations imposed by [*neighboring city/county*] to be fair to all parties, and I humbly suggest their adoption by our HOA, with assignment of oversight to a member or committee to review their effectiveness periodically.

If you have any further questions or concerns, please feel free to contact me directly.

Warm Regards,

[*Your name and contact information*]

WEIGHTS, MEASURES, *and* CONVERSIONS

INCHES	CM
⅛	0.3
⅓	0.8
½	1.3
¾	1.9
1	2.5
2	5.1
3	7.6
4	10
5	13
6	15
7	18
8	20
9	23
10	25
12	31
14	36

FEET	METERS
1	0.3
2	0.6
3	0.9
4	1.2
5	1.5
6	1.8
7	2.1
8	2.4
9	2.7
10	3.0
15	4.6
100	30.0

SQ. FT	SQ. M
2	0.19
3	0.30
4	0.37
25	2.32
32	2.97
50	4.65
2000	185.81

GALLON	LITER
1	3.8
2	7.6
3	11.4
5	18.9
12	45.4

TBSP.	ML
1	14.8
1 ½	22.2
2	29.6
4	59.1
10	147.9

TSP.	ML
¼	1.2
½	2.5
⅔	3.3
2	10.0

OUNCE	ML	G
1	29.6	28.34
4	118.3	113.4
5	147.9	141.7
6	177.4	170.1
8	236.6	226.8

CUP	ML
¼	60
½	120
1	240
2	480
5	1200

POUND	KG
1	0.5
1 ½	0.7
3	1.4
4	1.8
6	2.7
12	5.4
40	18.1
50	22.7
100	45.4
720	326.6

TEMPERATURES
$°C = \frac{5}{9} \times (°F - 32)$
$°F = (\frac{9}{5} \times °C) + 32$

ADDITIONAL RESOURCES

Although we set out to make this book as complete as possible, there is much more to be explored. In addition to consulting other texts and friends with chickens, you can find information about chickens and chicken keeping from several online forums, websites, books, and magazines.

Visit our website, **urbanfarmstore.com,** for more information and to explore the concepts, suppliers, and products mentioned in this book. We will also post additional information and updates to the book at this site.

FORUM

BackYard Chickens This is the largest and most active of many chicken-keeping forums. Want the hive-mind's take on your chicken's malady or your proposed coop design? This is the place for you. However, be prepared to receive conflicting and possibly inaccurate recommendations; it may be difficult to decide what advice to take. Still, there are not many places where you can post a photo related to chicken keeping and get a bunch of opinions of how to proceed. You can use these opinions as a starting point for your own independent research. BackYardChickens.com

COOPS

Omelet Eglu is the original mod coop for suburban chicken keepers. omlet.us/shop/chicken_keeping

Saltbox Designs, Seattle We love their coops! saltboxdesigns.wordpress.com

EM, PROBIOTICS, AND BOKASHI

Bokashicycle This is our favorite supplier for bokashi starter and supplies, with an extensive link section and original informational pages.
bokashicycle.com

EMRO This is the research organization that developed the original EM in Japan.
emrojapan.com

Teraganix Here you'll find official EM products, information, and recipes.
teraganix.com

FEED AND FEEDING

All About Feed This great site offers lots of nutritional data.
allaboutfeed.net

Feedipedia Ever wonder if grasshoppers make good feed ingredients? Check this online encyclopedia of animal feeds.
feedipedia.org

Ingredients 101.com Get nutritional data for many common and unusual feed ingredients here.
ingredients101.com

Silage "Small Scale Silage Production for Chicken Feed," by Troy Griepentrog, *Mother Earth News*.
motherearthnews.com/homesteading-and-livestock/
small-scale-silage-production-for-chicken-fee

Sprouted fodder "DIY Sprouted Fodder for Livestock," by Sarah Cuthill, *Mother Earth News.*
motherearthnews.com/homesteading-and-livestock/sprouted-fodder

HATCHERIES

. . . .

Greenfire Farms Visit their site, learn, and buy.
greenfirefarms.com

Ideal Poultry Breeding Farms This is one of the largest and most venerable hatcheries in the United States.
idealpoultry.com

Murray McMurray Hatchery This is another of the largest and most venerable hatcheries in the United States.
mcmurrayhatchery.com

HEALTH

. . . .

Backyard Poultry Medicine and Surgery: A Guide for Veterinary Practitioners, **edited by Cheryl B. Greenacre and Teresa Y. Morishita** This book is intended as a reference book for veterinarians. It is likely one of the first of many, as urban veterinarians realize the demand for chicken care. It is helpful for the layperson as well, although perhaps less accessible than books written for that audience.

The Chicken Health Handbook: A Complete Guide to Maximizing Flock Health and Dealing with Disease, **Second Edition, by Gail Damerow** This guide is thorough, easy to read, and provides clear instructions for simple surgical procedures for the very bravest chicken keepers. Highly recommended. Brief mentions of botanical treatments.

Merck Veterinary Manual This free online resource is a good starting point for getting basic information about poultry health. Many articles have been written specifically regarding backyard flocks.
merckvetmanual.com

National Center for Biotechnology Information (NCBI) database If you're looking for the original studies that have informed most of the health recommendations in this book, you'll find them by searching NCBI database. It contains hundreds of thousands of published peer-reviewed journal articles pertaining to veterinary care. Are you wondering whether that blog article recommending this or that treatment for your chicken's ailment is based on any data? You can look to see if there have been any studies done here.
ncbi.nlm.nih.gov

Poultry Health Handbook, **Fourth Edition, by L. Dwight Schwartz, DVM**
This online publication from a member of the faculty at Penn State College of Agriculture Sciences is a free and helpful resource for traditional veterinary approaches to poultry health care.
extension.psu.edu/publications/agrs-052

Poultry Hub This large site is aimed at scientists and farmers, but it's useful for everyone as a source of reliable information about chickens and their care.
poultryhub.org

UC Davis As a top research veterinary university in the United States, the UC Davis website is loaded with good stuff for chicken keepers.
vetmed.ucdavis.edu/index.cfm

NEEM
. . . .

Neem: A Tree for Solving Global Problems The National Research Council's 1992 book is available to view free of charge.
nap.edu/catalog/1924/neem-a-tree-for-solving-global-problems

Neem Foundation If you learned nothing else in this book, you now know how much we love neem! The Neem Foundation shares the love.
neemfoundation.org

PERMACULTURE

permacultureprinciples.com We have been deeply influenced by the principals of permaculture, a natural systems-based approach to energy and food production. Learn more general information here.
permacultureprinciples.com

Permies: Permaculture and Homesteading Goofballs This is a popular forum and information repository for permaculture and chicken-keeping enthusiasts.
permies.com

SUSTAINABLE AGRICULTURE

Environmental Protection Agency This is an excellent resource for general information about sustainability.
epa.gov/sustainability

National Center for Appropriate Technology NCAT supplies information, offers grants, and provides a free phone hotline for your questions through ATTRA, a national sustainable agriculture assistance program.
attra.ncat.org

URBAN FARMING

Mother Earth News This magazine and website have been invaluable resources for homesteaders and urban farmers since the dawn of time. The homesteading-lifestyle magazine is a classic, but the decade-spanning articles on the website are a free archive of accumulated wisdom and knowhow of the highest order. motherearthnews.com

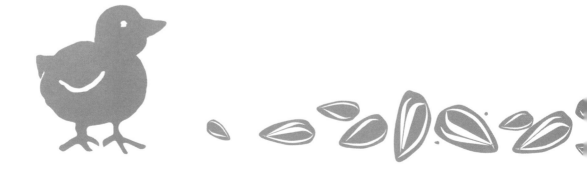

ACKNOWLEDGMENTS

We are deeply grateful for our loyal customers and readers who not only keep us in business, but trust us with their flocks and teach us so much. Thanks also to the many chicken bloggers and aficionados who have helped provide innovative solutions, photographs, and anecdotes for this book. Your enthusiasm for these quirky, engaging, and sometimes cuddly birds has reminded us why we became chicken keepers in the first place.

Thanks to Professor Cheryl Greenacre DVM, DABVP, whose speech at a veterinary conference opened our eyes both to the leading concerns of veterinarians who care for backyard birds and to the looming tsunami of the Veterinary Feed Directive. A special thanks to fellow Portlander Lisa Harrenstien, DVM, DACZM, who patiently consulted with us (and lent us Dr. Greenacre's book). A big thank you goes out to Gail Damerow for combing through our health chapter with a fine-toothed lice comb to snare numerous errors and omissions, both large and small.

We are forever indebted to our loyal and savvy literary agent, Carla Glasser, and particularly grateful that she introduced us to Stacee Lawrence at Portland's own Timber Press. We are also thankful that Stacee experienced a *Portlandia*-worthy moment when she overheard a conversation between passengers on a train about their chicken woes and realized there really are scads of perplexed chicken keepers out there in need of fresh answers. It was Stacee's calm demeanor and (nearly) limitless patience throughout this process that kept the project alive when an avalanche of snow days threatened to derail us. A tip of the cap also for Lisa Theobald, who asked the tough questions, kept us on course, and excelled at the difficult task of blending the work of two authors to produce a cohesive text. We bow in appreciation for the talents of the design team and everyone else at Timber Press who helped make this book not just informative, but beautiful as well.

Finally, we thank each other for keeping a sense of humor and abiding patience as we shoehorned several hundred hours of writing into our preternaturally hectic lives.

PHOTOGRAPHY AND ILLUSTRATION CREDITS

INDEX

© Paul Baker

ROBERT and **HANNAH LITT** are the founders of the Urban Farm Store in Portland, Oregon. They've been featured on Planet Green's *Renovation Nation*, National Public Radio, and Oregon Public Broadcasting, and Robert was named as one of Food & Wine's "40 Big Food Thinkers Under 40" in 2010.